DATE DUE			

POLYVINYLIDENE CHLORIDE

Polymer Monographs

Edited by HERBERT MORAWETZ, *Polytechnic Institute of New York*

A series of short monographs, each dealing with one specific polymer. They are written by experts from leading research laboratories, and cover equally basic scientific information on a polymer and information pertinent to its practical utilization.

POLYVINYLIDENE CHLORIDE

RITCHIE A. WESSLING

Dow Chemical Co.
Midland, Michigan

GORDON AND BREACH SCIENCE PUBLISHERS

New York   London   Paris

Copyright © 1977 by
   Gordon and Breach Science Publishers Inc.
   One Park Avenue
   New York, N.Y. 10016

Editorial office for the United Kingdom
   Gordon and Breach Science Publishers Ltd.
   41/42 William IV Street
   London W.C.2

Editorial office for France
   Gordon & Breach
   7-9 rue Emile Dubois
   Paris 75014

Library of Congress catalog card number 74–81400. ISBN 0 677
01700 6. All rights reserved. No part of this book may be reproduced or utilized in any form or by any means, electronic or
mechanical including photocopying, recording, or by any information storage and retrieval system,.without permission in writing from
the publishers. Printed in Great Britain.

Introduction to the Series

We hear much these days about the publication explosion and all of us are haunted by the specter of our inability to keep up with the technical literature. In this situation it becomes increasingly important to persuade competent and dedicated scientists to undertake the exacting task of summarizing the available knowledge on some given subject. Since the labor involved is very considerable, it has become the fashion to publish symposium volumes in which a number of experts contribute chapters on their speciality. However, this procedure leads to the publication of relatively expensive books which are bought almost exclusively by libraries and hardly at all by the individual investigator. This is obviously a very undesirable state of affairs, since nothing can replace the advantage gained from personally possessing literature which provides a background for our work.

This difficulty may be felt particularly strongly by those active in the polymer field. We have therefore decided to embark on the publication of a series of small books, modest in size and price, each summarizing present knowledge on one specific polymer. We believe that many who would hesitate in acquiring a large volume of some twenty chapters, of which only a few might be in the field of their own activity, will welcome the opportunity to select from the series those small books which will give the requisite background on polymers which specifically interest them.

HERBERT MORAWETZ

Preface

Vinylidene chloride copolymers of various types have now been
commercially available for over 35 years. In spite of the problems
associated with the handling of these materials, they continue to
find applications. In large part, this derives from the demand for
barrier coatings and films in the packaging area. As the complexity
of our food production chain has increased over the years, the
need to protect perishable products has grown also. This need can
be met in many cases by the utilization of vinylidene chloride
copolymers in package construction.

It was my good fortune to be able to discuss the origins and
development of Saran technology with many of the original
participants, particularly R. M. Wiley. My thanks to him and my
other Dow colleagues who contributed to this endeavor. I am
especially grateful to T. Alfrey, Jr., A. F. Burmester and D. R.
Carter for their helpful comments and suggestions, to Professor
H. Morawetz for reviewing the manuscript and finally to Mrs.
Sherry Tomczak for typing and correcting the many drafts,

RITCHIE A. WESSLING

Contents

CHAPTER 1

Introduction and Historical Aspects

Polyvinylidene chloride (PVDC) is the parent member of a family of polymers containing vinylidene chloride (VDC). Both the homopolymers and copolymers containing VDC as the major component are known as Saran.[†] The polymers used commercially are copolymers of VDC with various other unsaturated monomers; the homopolymer, sometimes referred to as Saran A, is primarily a laboratory curiosity. Dow has produced a variety of Saran polymers; the most common are copolymers with vinyl chloride known as Saran B resins.

PVDC has a number of outstanding characteristics. Since it is highly crystalline in the normal use temperature range (0-100°C), the polymer can be oriented into high strength fibers and films. It has outstanding resistance to solvents and chemical reagents other than bases. Of the organic coating materials, it has by far the lowest permeability to the widest variety of gases and vapors. However, these valuable properties cannot be easily exploited because of the polymer's extreme thermal instability. It cannot be heated to the molten state for longer than a few seconds without being severely degraded. In the process of decomposing, it eliminates large quantities of HCl, thereby further complicating any potential fabricating process.

A solution to the fabrication problem led to the commercial development of Saran polymers. The techniques, which included copolymerization, plasticization, stabilization, and new equipment design, were developed by Ralph Wiley and coworkers in the laboratories of The Dow Chemical Company during the period of

[†] A trademark of the Dow Chemical Company in most countries.

1933 to 1940. Commercialization of the first Saran resin was achieved toward the end of this period. After World War II, Saran grew rapidly into a major business. This growth was due in large part to the successful development of Saran film.

Because of their early R & D efforts, Dow achieved a dominant position in Saran. The initial series of patents resulting from the early work[1-18] grew to more than 200 by 1955.

A number of other companies besides Dow showed an early interest in Saran. I. G. Farben in Germany developed and commercialized copolymers of VDC with both vinyl chloride and ethyl acrylate.[19] They were used extensively in Germany during the war. Other companies active in VDC research in this period included du Pont in the U.S., Distillers, Ltd. in England, and BASF in Germany.

PVDC dates back to the very beginnings of polymer science. It was one of the polymers included in the pioneering research of Staudinger and coworkers. PVDC was first obtained in Dow in 1933 as a by-product from a chlorination reaction. A white precipitate formed in the low boiling fraction. It was identified by Reilley and Wiley as PVDC and, intrigued by its interesting properties, they began the line of research which led to the development of Saran.[20] PVDC was not unique to Dow, however. Many chemists starting with Renault in 1838 had prepared 1,1-dihaloethylenes and noted the spontaneous polymerization.[21] The product was not identified as PVDC until the investigations of Staudinger and Feisst[22] which were carried out in the late twenties. The initial attempts to utilize PVDC were frustrated by its insolubility and thermal instability. The Dow researchers set out to overcome these problems. Wiley[1] discovered that copolymers of VDC with vinyl chloride were less susceptible to degradation. When plasticized, they could be fabricated by the conventional methods of extrusion and molding. With the proper level of comonomer and plasticizer, a useful polymer could be obtained without too great a loss in the valuable properties.

At the same time, Wiley[23] discovered the ability of Saran copolymers to supercool and cold draw into highly oriented films and fibers. The most valuable property of these polymers, low permeability, was also observed in this fruitful period. The major applications utilizing this and other properties were identified. These included fibers, biaxially oriented films, molded pipe, and lacquer coatings.

A wide variety of copolymer compositions were screened during the early days. Only three of these have survived commercially over the years: the Saran B resins already mentioned are used in film. Copolymers with acrylonitrile (Saran F) are used in lacquer coatings and copolymers of VDC with various combinations of acrylates and methacrylates are used in latexes.

Almost from the very beginning, it became common practice to refer to all high VDC content copolymers as PVDC. As a consequence, many of the properties described in the literature as belonging to the homopolymer were actually measured on copolymer specimens of unknown composition. In spite of new rules of nomenclature for polymers, the tendency still exists to describe Saran copolymers as PVDC or simply Saran without specifying the type and amount of comonomer. Therefore, much of the published literature on these polymers is of limited scientific value.

Very little of the early work at Dow was published except in the patent literature. But Reinhardt has summarized part of it in a review written in 1943.[21] PVDC has been the subject of more recent reviews by Staudinger,[24] Schildknecht,[19] Gabbett and Smith,[25] Talamini and Peggion,[26] Serdynsky[27,28] and Wessling and Edwards.[29]

REFERENCES

1. R. M. Wiley, U.S. 2,160,931, to the Dow Chemical Company (1939).
2. R. M. Wiley, U.S. 2,160,932, to the Dow Chemical Company (1939).
3. R. M. Wiley, U.S. 2,160,933, to the Dow Chemical Company (1939).
4. R. M. Wiley, U.S. 2,160,934, to the Dow Chemical Company (1939).
5. R. M. Wiley, U.S. 2,160,935, to the Dow Chemical Company (1939).
6. J. H. Reilly and C. R. Russell, U.S. 2,160,936, to the Dow Chemical Company (1939).
7. J. H. Reilly, U.S. 2,160,937, to the Dow Chemical Company (1939).
8. J. H. Reilly and R. M. Wiley, U.S. 2,160,938, to the Dow Chemical Company (1939).
9. R. C. Reinhardt, U.S. 2,160,939, to the Dow Chemical Company (1939).
10. E. C. Britton, C. W. Davis and F. L. Taylor, U.S. 2,160,940, to the Dow Chemical Company (1939).

11.   E. C. Britton, C. W. Davis and F. L. Taylor, U.S. 2,160,941, to the Dow
      Chemical Company (1939).
12.   E. C. Britton, C. W, Davis and F. L. Taylor, U.S. 2,160,942, to the Dow
      Chemical Company (1939).
13.   E. C. Britton and C. W. Davis, U.S. 2,160, 943, to the Dow Chemical
      Company (1939).
14.   G. H. Coleman, U.S. 2,160,944, to the Dow Chemical Company (1939).
15.   R. M. Wiley, U.S. 2,160,945, to the Dow Chemical Company (1939).
16.   E. C. Britton, U.S. 2,160,946, to the Dow Chemical Company (1939).
17.   R. C. Reinhardt, U.S. 2,160,947, to the Dow Chemical Company
      (1939).
18.   R. M. Wiley, F. N. Alquist and H. R. Slagh, U.S. 2,160,948, to the
      Dow Chemical Group (1939).
19.   C. E. Schildknecht, *Vinyl and Related Polymers*, J. Wiley and Sons,
      New York, Chapter VIII (1952).
20.   J. H. Reilly and R. M. Wiley, U.S. 2,160,903; U.S. 2,160,904, to the
      Dow Chemical Company (1939).
21.   R. C. Reinhardt, *Ind. Eng. Chem.* **35**, 422 (1943).
22.   H. Staudinger and W. Feisst, *Helv. Chim. Acta.,* **13**, 805 (1930).
23.   R. M. Wiley, U.S. 2,183,602; U.S. 2,233,442, to the Dow Chemical
      Company (1939).
24.   J. J. P. Staudinger, *Brit. Plastics,* **20**, 381 (1947).
25.   J. F. Gabbett and W. M. Smith, in G. E. Ham (ed.), *Copolymerization*,
      Interscience, New York, Chapter X (1964).
26.   G. Talamini and E. Peggion, in G. E. Ham (ed.), *Vinyl Polymerization*,
      Vol. 1, Part 1, M. Dekker, New York, Chapter V (1967).
27.   E. D. Serdynsky, in *Chlorine, Its Manufacture, Properties and Uses*,
      Rheinhold, New York (1967).
28.   E. D. Serdynsky, in H. F. Mark, E. Cernia and S. M. Atlas (eds.), *The
      Chemistry and Technology of Fibers*, Vol. 3, J. Wiley and Sons, New
      York, p. 303 (1968).
29.   R. A. Wessling and F. G. Edwards, *Encyclopedia of Polymer Science
      and Technology,* **14**, 540 (1971).

CHAPTER 2

The Monomer

2.1 PREPARATION

VDC was originally prepared by reacting 1,1,2-trichloroethane (TCE) with a strong base.[1] This is still the preferred method. A stochiometric amount of base is required for complete conversion as shown in Equation (2.1).

$$Cl-CH_2-CHCl_2 + NaOH \xrightarrow[90\text{-}100\,^{\circ}C]{H_2O} CH_2\!\!=\!\!CCl_2 + NaCl. \qquad (2.1)$$

When the reaction is carried out at atmospheric pressure, VDC distills out of the mixture as it forms. An efficient condenser is required to trap the highly volatile product.

Essentially the same method is used to make VDC commercially.[2] TCE which is obtained by chlorination of vinyl chloride is reacted with lime or caustic at 98-99°C. The process is carried out continuously with a slight excess of base giving yields of 90% VDC. The crude product is washed with water and dried before fractionally distilling.

An inhibitor is added at this point to prevent oxidation or polymerization of the monomer. A number of compounds have been found to be effective.[3-6] The types used commercially are phenol and p-methoxyphenol.[7] A typical analysis of a commercial product is shown in Table 2.1.

Vinylidene chloride can be made from other starting materials. Among those reported to give good yields are bromochloroethane, trichloroethyl acetate, tetrachloroethane and acetylene. There has been an interest in developing a catalytic process for VDC but none

TABLE 2.1

Typical Analysis of Vinylidene Chloride[7]

Vinylidene chloride (excluding inhibitors), wt. %	99.7
Vinyl chloride, ppm	850
cis-1,2-dichloroethylene, ppm	500
trans-1,2-dichloroethylene, ppm	1500
1,1-dichloroethane, ppm	10
Ethylene dichloride, ppm	10
1,1,1-trichloroethane, ppm	150
Trichloroethylene, ppm	10
Inhibitor	
MEHQ grade, ppm	200
Phenol grade, wt %	0.6-0.8

has reached commercial status. Processes such as thermal cracking of TCE[8] and chlorination of acetylene[9] have been described.

The only significant use for vinylidene chloride is in the manufacture of various copolymers; MEHQ inhibited monomer can normally be used for this purpose without purification. All that is required to effect polymerization is a slightly higher initiator level. Phenol, on the other hand, interferes too much with polymerization when present at the levels shown in Table 2.1. It can be removed either by an alkaline wash or by distillation of the monomer. Even when stabilized, the monomer should be kept cold and under an inert atmosphere where possible. It is preferably handled in glass, phenolic or Teflon lined metal equipment, though stainless steel and nickel are adequate. Contact with iron, copper and aluminium must be avoided.[7]

2.2 PROPERTIES

VDC is a colorless, sweet smelling, highly volatile liquid at ambient temperatures. Considering its commercial importance, relatively few engineering data are available, but Gallant[10] has estimated some of the more important physical properties. The Dow Chemical Company monomer bulletin is another source of physical property data (Table 2.2).

The structure of VDC has been studied both by electron

TABLE 2.2

Physical Properties of Vinylidene Chloride Monomer[7]

Property	Value
Molecular weight	96.95
Boiling point at 760 mmHg, $^{\circ}$C	31.56
Freezing point, $^{\circ}$C	− 122.5
Density	
at −20 $^{\circ}$C	1.2902
at 0 $^{\circ}$C	1.2517
at 20 $^{\circ}$C	1.2132
Refractive index	
at 10 $^{\circ}$C	1.43062
at 15 $^{\circ}$C	1.42777
at 20 $^{\circ}$C	1.42468
Latent heat of vaporization, cal/mole	
at 25 $^{\circ}$C	6328
at boiling point	6257
Dielectric constant, 16 $^{\circ}$C	4.67
Vapor pressure, mmHg (T in $^{\circ}$C)	
log Pmm = 6.98200-1104.29/(T + 237.697)	
Specific heat, cal/g	0.275
Heat of combustion, kcal/mole	261.93
Heat of polymerization, kcal/mole	−18.0 ± 0.9
Absolute viscosity, cps	
at −20°C	0.4478
at 0°C	0.3939
at 20°C	0.3302
Flash point, $^{\circ}$F	
Tag closed cup	55
Cleveland open cup	5
Explosive mixture with air, % by vol.	
Lower limit	7.3
Upper limit	16.0
Auto-ignition temperature, $^{\circ}$F	1031
Solubility of monomer in H_2O, wt. % (25 $^{\circ}$C)	0.021
Solubility of H_2O in monomer, wt. % (25 $^{\circ}$C)	0.035

diffraction[11] and microwave spectroscopy.[12] According to Evans,[13] the latter results are consistent with other thermal and spectroscopic studies. The molecule is planar as expected with the bond angles and distances listed in Table 2.3.

TABLE 2.3

Structure of Vinylidene Chloride

Bond	Distance or angle	Reference
C–H	1.07 Å	12
C–C	1.32 Å	12
C–Cl	1.727 Å	12
∠H–C–C	120°	12
∠C–C–Cl	123° 10′	12
∠Cl C–Cl	114.5°	11
C–Cl	1.71 Å	11
∠C–C–Cl	122.8°	11

Stull and coworkers have made an extensive study of the thermal properties of VDC.[14,15] They determined melting point, boiling point and heats of transition. They also reported heat capacity and vapor pressure as a function of temperature. These data were used to calculate the thermodynamic properties of VDC.

Evans[13] analyzed the spectroscopic data on VDC. He compared his results from both IR and Raman spectra to earlier studies.[16,17]

Using the microwave structure to calculate moments of inertia, Evans found the spectroscopic properties to be in agreement with the calculated ideal gas behavior reported by Stull *et al*.

Vinylidene chloride is miscible with most organic solvents and monomers. It is only slightly soluble in water at 25 °C and atmospheric pressure, though the exact level is uncertain. The Dow Chemical Company monomer bulletin gives a value of 0.0022 moles/liter (0.021 wt. %).

TABLE 2.4

Solubilities and VDC-Soap Ratios at 25 °C[18]

Potassium laurate concentration (mole/1)	0.0	0.05	0.10	0.20
Solubility of VDC (moles/1)	0.066	0.079	0.098	0.144
Mole VDC per mole soap at saturation	–	0.26	0.32	0.39

The value reported by Wiener[18] is substantially higher. He studied the solubility of VDC both in H_2O and in soap solutions. His results are given in Table 2.4.

2.3 CHEMISTRY

The double bond in VDC is quite reactive and undergoes the usual addition reactions characteristic of olefins. Only two reactions are of interest here: polymerization and oxidation. The latter is important in the handling and use of VDC.

The pure monomer is stable in the dark, but exposure to oxygen causes it to become turbid and develop a sharp acrid odor. The turbidity is due to minute PVDC crystals (PVDC is insoluble in the monomer) and a polymeric peroxide which may have the structure:

$$\left(CH_2 - CCl_2 - O - O \right)_n$$

The peroxides decompose to phosgene and formaldehyde, hence the odor.

Mayumi et al.[19,20] suggest that the peroxide has the cyclic structure

$$\begin{array}{cc} CH_2 - CCl_2 \\ | \qquad | \\ O \; - \; O \end{array}$$

They isolated this compound and showed that it could be mechanically exploded. It was used to initiate the polymerization of VDC.[19]

The process by which phosgene is formed is not well defined. But it probably arises from decomposition of an intermediate radical species. Autoxidation of chlorinated ethylenes is well known. The mechanism suggested by Walling[21] is probably applicable to VDC also. Under some conditions, the reaction can yield chloroacetyl chloride.[22]

Vinylidene chloride polymerizes most readily by free radical initiation. It does not polymerize easily by a cationic mechanism, but can be made to polymerize anionically. It copolymerizes readily with a wide variety of monomers. These aspects are discussed fully in Chapter 3.

The polymerization chemistry of VDC can be better understood by examining the relationship between its structure and reactivity. The reactivity of a monomer is determined by the nature of the substituents on the double bond.[23-26] The influence of these substituents on polymerization kinetics can be discussed in terms of resonance, polarity, and steric interactions.[27]

One way of looking at reactivity is by comparing thermodynamic changes. The Gibbs free energy change, ΔG_p, for the polymerization process is a measure of the driving force towards polymerization. Although one cannot draw conclusions about kinetics from a knowledge only of the thermodynamic changes involved in a reaction, Joshi and Zwolinski[28] observe that there often is a correlation between the two for series of similar compounds undergoing the same reaction. They also point out that the entropy of polymerization, ΔS_p, is approximarely the same for all addition polymerizations. The critical factor governing polymerizability is the change in enthalpy, ΔH_p.

Both ΔH_p,[14,29,30] and ΔS_p,[31] have been determined for the polymerization of VDC. The experimental values when corrected for the partial crystallinity of PVDC [$\Delta H_f = 1.5$ kcal/mole[32]] indicate that addition is sterically hindered. Even so, the monomer does polymerize easily and falls about in the middle on a thermodynamic scale of reactivity.[28]

Reactivity can also be analyzed in terms of the Q-e scheme provided the steric factors are not dominant.[27] This appears to be the case for VDC. Its ceiling temperature (where monomer is in equilibrium with polymer) is $\sim 350\ °C$,[31] well above the polymerization temperature range. The only abnormal case of copolymerization is with isobutylene where the simple copolymerization theory has been shown to be inadequate to explain the results.[33]

The Q-e scheme is an empirical method of correlating monomer reactivities in copolymerization. The effect of resonance on monomer reactivity is characterized by Q; the relative polarity of the radical is described by e. These quantities can be related to the monomer reactivity ratios according to the following equations:

$$r_1 = Q_1/Q_2 \exp[-e_1(e_1 - e_2)] \tag{2.2}$$

$$r_2 = Q_2/Q_1 \exp[-e_2(e_2 - e_1)] \tag{2.3}$$

Reactivity ratios are discussed further in Chapter 3.

The Q-e map (plot of e vs log Q)[34] is an illustrative method of displaying monomer reactivities. VDC falls just about in the middle of the map, indicating that the double bond is mildly stabilized by resonance with the chlorine substituents and is also polarized due to their electronegative character leading to a positive value of e. According to the Q-e scheme, VDC should be easily copolymerized by a free radical mechanism with monomers of similar character such as methyl acrylate. This is clearly in agreement with observation. It should not polymerize readily with resonance stabilized monomers or with monomers containing strongly electron-deficient or electron-rich double bonds.

An analysis in terms of the Q-e scheme also indicates that VDC should polymerize or copolymerize more readily by an anionic mechanism than a cationic mechanism.[35]

REFERENCES

1. C. E. Schildknecht, *Vinyl and Related Polymers*, Wiley, N.Y., Chap. VIII (1952).
2. P. W. Sherwood, *Ind. Eng. Chem.*, **54**, 29 (1962).
3. E. C. Britton and W. J. LeFevre, U.S. 2,121,009; U.S. 2,121,010; U.S. 2,121,011; U.S. 2,121,012, to the Dow Chemical Company (1938).
4. G. H. Colman and J. W. Ziemba, U.S. 2,136,333; U.S. 2,136,334, to the Dow Chemical Company (1938).
5. G. H. Coleman and J. W. Ziemba, U.S. 2,160,944, to the Dow Chemical Company (1939).
6. R. M. Wiley, U.S. 2,136,347; U.S. 2,136,348; U.S. 2,136,349, to the Dow Chemical Company (1938).
7. *Vinylidene Chloride Monomer*, Product Bulletin, Dow Chemical Company (1968).
8. A. W. Hansen and W. C. Goggin, U.S. 2,238,020, to the Dow Chemical Company (1941).
9. A. Jacobowsky *et al.*, U.S. 2,915,565, to Knapsack-Griesheim A.C. (1959).
10. R. W. Gallant, *Hydrocarbon Processing*, **45**, 153 (1966).
11. R. L. Livingston, C. N. Ramachandra Rao, L. H. Kaplan and L. Rocks, *J. Am. Chem. Soc.*, **80**, 5368 (1958).
12. S. Sekino and T. Nishikawa, *J. Phys. Soc., Japan*, **12**, 43 (1957).
13. J. C. Evans, *J. Chem. Phys.*, **30**, 934 (1959).
14. G. C. Sinke and D. R. Stull, *J. Phys. Chem.*, **62**, 397 (1958).

15. D. L. Hildebrand, R. A. McDonald, W. R. Kramer and D. R. Stull, *J. Chem. Phys.,* **30**, 930 (1959).
16. H. W. Thompson and P. Torkington, *Proc. R. Soc.,* **A184**, 21 (1954).
17. P. Joyner and G. Glockler, *J. Chem. Phys.,* **20**, 302 (1952).
18. H. Wiener, *J. Polym. Sci.,* **7**, 1 (1950).
19. K. Mayumi, *Jap.,* 5040 (1950).
20. K. Mayumi, O. Shibuya and S. Jikinose, *Nippon Kagaku Zasshi,* **78**, 280 (1957) CA 53: 5225a.
21. C. Walling, *Free Radicals in Solution,* Wiley and Sons, N.Y. (1957).
22. A. Rieche and H. Stetter, Ger. 746,451 (1941).
23. P. J. Flory, *Principles of Polymer Chemistry,* Cornell Univ. Press, Ithaca, N. Y. (1953).
24. G. M. Burnett, *Mechanism of Polymer Reactions,* Interscience, N.Y., (1954).
25. C. H. Bamford, W. G. Barb, A. D. Jenkins and P. F. Onyon, *The Kinetics of Vinyl Polymerization,* Butterworth, London (1958).
26. G. E. Ham (ed.), *Vinyl Polymerization,* Vols. I and II, M. Dekker, New York (1967).
27. T. Alfrey, Jr., and L. J. Young, in G. E. Ham (ed.), *Copolymerization,* Interscience, New York, Chapter II (1964).
28. R. M. Joshi and B. J. Zwolinski, in G. E. Ham (ed.), *Vinyl Polymerization,* Vol. I, M. Dekker, New York (1967).
29. L. K. J. Tong and W. O. Kenyon, *J. Am. Chem. Soc.,* **69**, 2245 (1947).
30. R. M. Joshi, *Makromol. Chem.,* **55**, 35 (1962).
31. B. V. Lebedev, I. B. Rabinovitch and V. A. Budarina, *Vysokol. soyed,* **A9**, 488 (1967).
32. K. Okuda, *J. Polym. Sci.,* **A2**, 1749 (1964).
33. J. B. Kinsinger, T. Fischer and C. W. Wilson, *Polym. Lett.,* **5**, 285 (1967).
34. T. Alfrey, Jr., J. J. Bohrer and H. Mark, *Copolymerization,* Interscience, N.Y. (1952).
35. A. Konishi, *Bull. Chem. Soc., Japan,* **35**, 193 (1962).

CHAPTER 3

Polymerization

3.1 HOMOPOLYMERIZATION

3.1.1 Homopolymerization by free radical initiation

Vinylidene chloride (VDC) homopolymerizes by both free radical and ionic mechanisms. In most respects, it behaves like a typical vinyl monomer.[1] The abnormal features of its polymerization chemistry are a consequence of the fact that PVDC does not dissolve in its monomer. Consequently, it separates as another phase from the polymerization reaction as it is formed. The resulting heterogeneity exerts a marked influence on the kinetics of the reaction. (The kinetics of heterogeneous polymerization are examined in Chapter 4).

The subject of VDC polymerization has recently been reviewed by Talamini and Peggion.[2] A number of older reviews are still of interest also.[3-6] Most of the published work relates to free radical polymerization, and has been primarily of a practical nature. Not unexpectedly, the volume of patent literature is large. Regretfully, this mass of literature adds little to our understanding and is, therefore, not critically reviewed in this book.

The homogeneous polymerization of VDC in a medium dissolving both monomer and polymer has not been reported. The free radically initiated reactions are carried out in mass or solvent slurry, suspension and emulsion. The original Saran process was based on a mass reaction, which is very hazardous on a large scale. It is strongly autocatalytic and exothermic and requires efficient stirring and heat transfer to be kept under control. This is very difficult to achieve because of the heterogeneity of the reaction. The precipitat-

13

ing polymer causes the reaction mixture to thicken at very low conversions. By 10-20% conversion, it sets up into a thick paste. Further reaction converts it to a hard, porous, bonelike mass. The conversion-time curve for a benzoyl peroxide initiated reaction is shown in Figure 3.1. During the final stage, hot spots can develop in the reaction due to the high rate and poor heat transfer. The reaction can be controlled by stopping at about 30% conversion and stripping off the unreacted monomer. If it is allowed to continue, the heat buildup causes the polymer to decompose releasing large quantities of HCl. This can easily lead to an explosion. For these reasons, the process was soon switched to a dispersed system in which stirring could be maintained.[7]

The mass polymerization is still a useful method at the laboratory level. Reactions in small tubes can be carried to completion without difficulty. Polymers of very high molecular weight can be prepared in this way. A fine, highly crystalline powder of PVDC can be obtained by stopping the reaction in the paste stage and flashing off the unreacted monomer. Alternatively, the reaction can be carried out in an inert diluent such as cyclohexane or benzene. If the initial monomer concentration is less than about 50%, the reaction can be carried to completion without obtaining a solid mass. The obvious disadvantage of a solvent system is that the polymer must be freed of solvent before using. In addition, molecular weights and rates of conversion are lower.

Free radical polymerizations can be initiated both by chemical initiators and by radiation. Purely thermal initiation occurs[8] but rates are too low for this to be a practical method. The spontaneous polymerization of VDC so often observed at room temperature is due either to peroxides formed by exposure to oxygen or by exposure to sunlight. Very pure monomer will not polymerize in the dark at low temperatures. A list of some of the initiators that have been found useful is given in Table 3.1.

Photochemical initiation is another method which induces free radical polymerization of VDC. The method has the advantage of not requiring chemical initiators or promoters. It can also be used at low temperatures. The method has been studied in detail by Burnett and Melville.[11] They reported that the rate is sensitive to the wave length of the radiation employed. The highest overall rate was

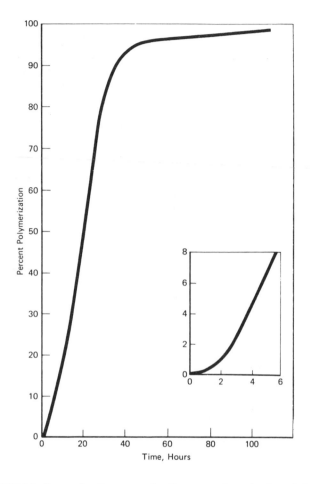

FIGURE 3.1 Conversion-time curve for the mass polymerization of vinylidene chloride (Ref. 6). 0.5% benzoyl peroxide; 45 °C.

obtained with a broad spectrum UV lamp.

The utility of photoinitiation is severely limited by two undesirable features. As polymerization proceeds, the turbid reaction medium scatters more of the incident light. This produces a significant diminution in the rate of initiation.[12] The other undesirable

TABLE 3.1

Chemical Initiators for the Free Radical Polymerization of VDC in Mass
or Slurry

Initiator	Use range, °C	Reference
Benzoylperoxide	40 to 80	5
Azobisisobutyronitrile	35 to 70	4
Isopropylpercarbonate	0 to 40	9
Tri-n-butyl boron/O_2	−78 to 25	10

side effect is the photochemically induced decomposition of the
polymer. When PVDC is exposed to strong UV radiation it may
eliminate HCl. Prolonged exposure causes discoloration of the
polymer.[13] The extent of degradation can be limited by polymerizing at low temperature.

Other forms of high energy radiation such as γ-rays will also
polymerize VDC.[8] This appears to be a radical reaction under most
conditions. For example, Burlant and Green demonstrated the
radical nature of this process by analyzing copolymerization data.[14]
On the other hand, at temperatures below 0 °C there is a possibility
that some ionic polymerization occurs in mixtures with isobutylene.[15,16] In still another study, γ-ray initiated emulsion polymerization occurred in the presence of water[17] which suggests a radical
initiation process.

VDC has been polymerized in the solid state by irradiating a
thiourea canal complex.[18] Polymerization of the pure crystalline
monomer has apparently not been successful.[19] VDC has not been
observed to polymerize in the vapor phase either. Its volatility and
ability to absorb radiation would suggest this as a likely possibility.
Irradiation with x-rays, however, induced polymerization only in the
condensed or adsorbed phase, the vapor remaining unaffected.[20]

PVDC can also be prepared in aqueous emulsion[21,22] or suspension.[23] These methods are useful for preparing large quantities of
polymer. But they are complicated processes both from the
theoretical and the practical point of view. The reactions are doubly
heterogeneous since monomer and polymer are incompatible and

both are insoluble in water. The recipes contain a number of ingredients that must be removed before the polymer can be used.

Conversion-time curves for the emulsion polymerization of VDC are shown in Figure 3.2.

The reaction is carried out in a stirred kettle equipped with an efficient condenser. The ingredients are added and the mixture brought to refluxing temperature, $\sim 32\,^\circ$C; efficient stirring must be maintained during the reaction.

100 parts	VDC
100 parts	H_2O
1.0 parts	sodium lauryl sulfate
0.15 parts	ammonium persulfate
0.15 parts	sodium bisulfite

The reaction is complete in 5-6 hours. The polymer can be recovered by coagulating the latex either by freezing or adding an electrolyte. It should be washed thoroughly to remove ionic contaminants.

Redox systems employing water-soluble initiators are normally used in emulsion polymerization. Oil-soluble initiators have been used but the rates are lower. A list of possible Redox pairs is given in Table 3.2. VDC should be polymerized at pH < 9 because the polymer is attacked by aqueous base.

TABLE 3.2
Redox Initiators for the Emulsion Polymerization of VDC

Initiator	Activator
Ammonium persulfate	Sodium bisulfite
Hydrogen peroxide	Ferrous salts
Cumene hydroperoxide	Sodium formaldehyde – sulfoxylate
t-butyl hydroperoxide	Sodium formaldehyde – sulfoxylate
Potassium persulfate	Sodium formaldehyde – sulfoxylate
Isopropyl percarbonate	Sodium dithionate

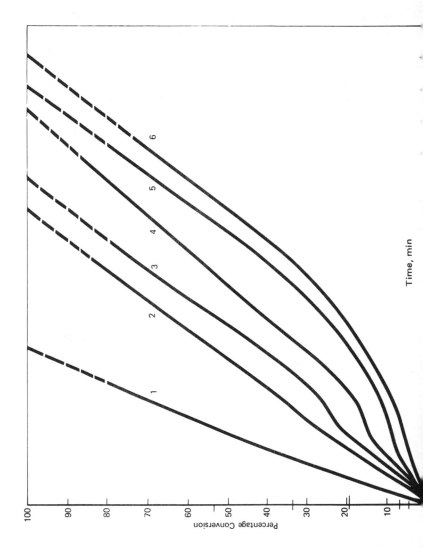

FIGURE 3.2 Conversion-time curves for the emulsion polymerization of VDC at 36.0 °C. Initial catalyst charge: 0.15 g $(NH_4)_2S_2O_8$, 0.15 g $Na_2S_2O_5/100$ g monomer. Sodium lauryl sulfate concentrations: (1) 10.0 g, (2) 5.0 g, (3) 3.0 g, (4) 2.0 g, (5) 1.0 g, (6) 0.50 g/100 g monomer (Ref. 22).

The advantages of emulsion polymerization are that high molecular weights can be achieved with high but controlled rates of polymerization. Suspension polymerization has the same potential for control because of the fluidity of the aqueous suspending medium but the rates required to achieve high molecular weights are lower.

A typical recipe for the suspension polymerization of VDC is given below:

100	parts	VDC
300	parts	water
0.05	parts	methyl hydroxypropyl cellulose
0.3	parts	lauryl peroxide

The ingredients are emulsified in a suitable colloid mill and polymerized under pressure at 50 °C for 30 h or more. The polymer is obtained by filtering and washing. The suspension process, though slower, requires less cleanup to recover pure polymer.

3.1.2 Other methods of polymerization

The ionic polymerization of VDC has received relatively little attention. This is primarily because the polymer structure is not affected by the polymerization mechanism — in contrast to polyolefins. Another deterrent to the use of ionic polymerization is that many of the initiators also catalyze the decomposition of the polymer.

As pointed out earlier, cationic polymerization may occur in low temperature γ-ray initiated copolymerizations, but only one clear-cut example of cationic homopolymerization has been reported (Zilberman and coworkers[24] polymerized VDC with $ZnCl_2$ at 40 °C. They obtained both the dimer and a low molecular weight polymer.

An anionic reaction has also been observed. Konishi reports that VDC polymerizes in hexane in the presence of butyl lithium.[25] The products were of low molecular weight and were lower in chlorine and more soluble than pure PVDC. These observations suggest that either dehydrochlorination or alkylation of the polymer took place. The anionic mechanism was established by lack of inhibition with known radical scavengers and by copolymerization studies.

Erusalimskii[26] and coworkers showed that the efficiency of the BuLi catalyst could be improved markedly by forming complex mixtures with butyl iodide, or tributyl dimagnesium iodide. While no details of polymerization of VDC were given, undegraded high polymers of VC and chloroprene were obtained.

Co-ordination catalysis has been reported recently to effect the copolymerization of VDC.[27] The catalyst was a $C_2 H_5 AlCl_2$ / $Ti(-O-C_4H_9)_4$ complex. The reactivity ratios in copolymerizing VDC with VC indicated that this was not a case of free radical growth. Like metal alkyls, the Ziegler catalysts tend to produce partially degraded products. A British patent[28] describes how the degradation can be avoided by using certain coreactants. The catalyst system contains three components.

1) A transition metal salt such as $TiCl_3$.
2) An organo aluminium compound such as $Al(CH_3)_3$.
3) An aliphatic alcohol of 2 to 10 carbon atoms containing an amino, cyano or halogen substituent such as ethylene chloro-hydrin.

The catalyst is preferable prepared by reacting components (2) and (3) in a 1 : 1 mole ratio and adding the product to the metal salt in a mole ratio of about 1 : 1 to 3 : 1. The polymerization can be carried out under an inert atmosphere either neat or in a chlorinated solvent, preferably in the temperature range of 30-70 °C. A typical recipe is given below:

1) 5.0 millimoles of adduct of triethylaluminum and ethanolamine (1 : 1)
2) 2.5 millimoles $VOCl_3$
3) 10 ml VDC (0.125 moles)
4) 70 ml 1,2-dichloroethane (solvent)

The solvent was introduced into a pressure vessel and purged with an inert gas. The reactants were added in order. The vessel was sealed and heated at 60 °C for 13 h. The products were precipitated with methanol. Yield 59.1%. The IR spectrum was the same as that of a free radical polymer.

   While ionic polymerization of VDC has been of little value in the past, the use of Ziegler catalysts to effect polymerization may change this picture. The motivation would be to prepare copolymers that are inaccessible by free radical processes. In the present state of the art, the various methods of polymerizing VDC are listed in Table 3.3. The free radical polymerizations in suspension and emulsion must be considered the only ones of general utility until more information on the new approaches is available.

TABLE 3.3
Methods for Polymerizing VDC

Mechanism	Medium or method	Scope
Radical	Mass	Lab method
Radical	Slurry	Lab method
Radical	Emulsion (Aq)	General utility
Radical	Suspension (Aq)	General utility
Radical	Photochemical	Limited value
Radical	High energy radiation	Limited value
Ionic	Slurry	Limited value
Co-ordination complex	Slurry	Unknown

3.2   COPOLYMERIZATION

VDC can be polymerized with a wide variety of monomers using either free radical or ionic initiation. The number of possible combinations is enormous, only a limited number of which have been prepared. The objective in most cases has been to prepare a material with specific properties rather than to study the copolymerization of VDC. Design of commercial polymers is discussed in Chapter 11.

   The early scientific investigations of the copolymerization of VDC were closely associated with efforts to test the copolymer composition equation.[29,30] This equation was derived for a homogeneous batch copolymerization. It was assumed that the reactivity of chain ends was unaffected by adjacent units (no penultimate effect). In this model, there are four possible elementary reactions:

$$M_1^{\cdot} + M_1 \xrightarrow{k_{11}} M_1^{\cdot}$$

$$M_1^{\cdot} + M_2 \xrightarrow{k_{12}} M_2^{\cdot}$$

$$M_2^{\cdot} + M_1 \xrightarrow{k_{21}} M_1^{\cdot}$$

$$M_2^{\cdot} + M_2 \xrightarrow{k_{22}} M_2^{\cdot}$$

where $M_1^{\cdot}$ and $M_2^{\cdot}$ denote chain radicals with a terminal unit of monomers 1 and 2, respectively.

A steady-state analysis leads to the following relationship between instantaneous copolymer composition and monomer composition

$$\frac{m_1}{m_2} = \frac{dM_1}{dM_2} = \frac{M_1}{M_2}\left[\frac{r_1 M_1 + M_2}{r_2 M_2 + M_1}\right],$$

where m_1 are moles of each monomer entering the copolymer and r_i are the reactivity ratios defined by

$$r_1 = k_{11}/k_{12}; \quad r_2 = k_{22}/k_{21}.$$

Many studies were carried out to determine reactivity ratios for various monomer pairs. VDC was one of the four monomers included in the significant study by Lewis, Mayo and Hulse.[31] (The others were styrene, acrylonitrile, and methyl methacrylate.) The reactivity of these monomers in all possible combinations led them to suggest the existence of a general order of reactivity. This in turn stimulated the development of the Q-e scheme by Alfrey and Price.[32]

These and other results indicated that VDC acts as a typical vinyl monomer in copolymerization. Steric effects like those which complicate the polymerization of 1,2 dichloroethylenes have been observed in only one case, the copolymerization with isobutylene.[33] The only abnormal characteristic observed in the copolymerization of VDC is the tendency for polymers with high VDC content to precipitate from the reaction mixture.

3.2.1   Free radical copolymerization

As in the case of homopolymerization, the free radical copolymerization of VDC has been more widely studied. It copolymerizes
by this mechanism most readily with acrylates. The reactivity
ratios for many of these pairs are close to unity.[34] Copolymerization either with very unreactive monomers like vinyl acetate or with
monomers like styrene that form stable radicals takes place with
greater difficulty. This is in accord with the predictions of the Q-e
scheme.[29] (The Q-e scheme is discussed in Chapter 2.) Some
comparisons are shown in Table 3.4.

TABLE 3.4

The Reactivity of Vinylidene Chloride with Important Monomers

Monomer	Q^a	e	r_1(VDC)	r_2
Vinylidene chloride (VDC)	0.22	0.36	–	–
Styrene (STY)	1.00	−0.80	0.14	2.0
Butadiene (BD)	2.39	−1.05	<0.05	1.9
Vinylchloride (VC)	0.044	0.20	3.25	0.30
Maleic anhydride (MAH)	0.23	2.25	9.0	0.0
Acrylonitrile (AN)	0.60	1.20	0.49	1.20
Butylacrylate (BA)	0.50	1.06	0.88	0.83
Methylacrylate (MA)	0.42	0.60	1.0	1.0
Methyl methacrylate (MMA)	0.74	0.40	0.24	2.53
Vinyl acetate (VAC)	0.026	−0.22	5.0	0.05
Ethyl vinyl ether (EVE)	0.032	−1.17	3.2	0.0

[a] Q-e values from Ref. 35.

A more complete listing of reactivity ratios is available in the
Polymer Handbook.[35] Given the reactivity ratios for a monomer
pair, the instantaneous copolymer composition formed in any
monomer mixture can be calculated

$$F_1 = \frac{r_1 f_1{}^2 + f_1 f_2}{r_1 f_1{}^2 + 2f_1 f_2 + r_2 f_2{}^2}$$

where F_1 is mole fraction of species 1 in the copolymer and f_1 is
the mole fraction of species 1 in the monomer. Instantaneous
copolymer composition plots for some of the important Saran

copolymers are shown in Figure 3.3. The values for r_1 and r_2 selected for these plots are probably the best values. There is a considerable variation in values reported for the most widely studied systems, VDC/VC and VDC/acrylonitrile. This is illustrated for the latter pair by the data in Table 3.5.

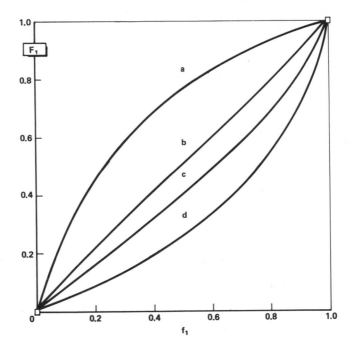

FIGURE 3.3 Instantaneous copolymer composition plots for important systems. f_1 = mole fraction VDC in feed; F_1 = mole fraction VDC in copolymer.
a: VDC/vinyl chloride; b: VDC/butyl acrylate; c: VDC/methyl methacrylate; d: VDC/acrylonitrile.

(Courtesy V. E. Meyer, The Dow Chemical Company, Midland, Michigan.)

TABLE 3.5
Effect of Conversion on Calculated Reactivity Ratios
for the System VDC/AN

Medium	$T\,^{\circ}C$	% Conversion	r_1	r_2	Ref.
Mass	60	15-43	0.37	0.91	31
40 wt. % monomers in					
t-butanol	43-47	1.5-7.8	0.49[a]	1.20[a]	36
25 wt. % monomers in					
t-butanol	50	84-95	~1	~1	37
Emulsion	36		0.39	0.59	38

[a] Probable best values.

The heterogeneity of the reaction may have influenced the values reported. This is more clearly shown by copolymerization in emulsion. The apparent reactivity ratios depend on surfactant concentration.[39] When this is not taken into account, abnormal values of r_1 and r_2 are obtained.

The measurement of reactivity ratios has continued to interest polymer chemists over the years. In some cases, they are measured in the process of evaluating new monomers. For example, Marvel et al. studied the copolymerization characteristics of the vinyl ketostearates.[40] VDC was one of the monomers used in the study. Recently, Winston and Li[41] used VDC as one of the monomers in evaluating the reactivity of 4-cyclopentene-1,3-dione. Many times, however, the purpose of a copolymerization study has been to re-examine previously investigated systems using improved methods of analyzing copolymer composition,[42] and techniques for calculating reactivity ratios.[43]

The accurate determination of reactivity ratios is notoriously difficult.[29] In the classical approach, compositions of low conversion copolymers prepared at various monomer ratios were determined by elemental analysis. Reactivity ratios were obtained from the raw data by graphical procedures. The method was prone to serious error due to low accuracy of the analyses and to composition drift.

The analytical problem is especially severe for vinylidene

chloride/vinyl chloride copolymers.[44] The calculated compositions are sensitive to small differences in measured chlorine content. In the composition range of interest, 75-95% VDC, the total difference is only $\sim 3\%$ (PVDC 73.14% Cl; PVC $-$ 56.8% Cl). This problem is magnified by the fact that theoretical chlorine contents are rarely obtained in PVDC; values of 72-73% Cl are commonly observed. Chlorine is normally measured by Parr bomb oxidation with Na_2O_2 followed by $AgNO_3$ titration of the chloride ion.[45] Nondestructive methods include x-ray fluorescence and neutron activation analysis.

Other methods of determining copolymer composition include NMR, IR and Raman spectroscopy and pyrolysis-gas chromatography. NMR has been used most extensively, particularly for VDC/VC copolymers,[42,46-48] but for other systems as well.[33,42,49-51,52-55] Infrared methods are useful only in certain cases.[55-60] Both the Raman method[61] and the pyrolysis technique[62-63] are quite new and have been applied only to the VDC/VC system.

Compositions must be measured on very low conversion polymers in order to minimize errors from composition drift. In any copolymerization, the polymer formed initially is richer in the more reactive component; but as the supply of this species is depleted, the polymer incorporates more of the less reactive monomer. The composition drift can be calculated from the integrated form of the Skeist equation.[64] A number of computer programs have been written to carry out this calculation.[65,66] A typical computer printout of composition vs. conversion for the VDC/VC copolymer systems is shown in Figure 3.4. When $r_1 = r_2 = 1$ as in the case of VDC/MA, no drift occurs. This is very nearly the case for n-butyl acrylate also. Composition drift in the AN system is significant but not nearly as severe as cases where VDC is copolymerized with a much less reactive comonomer like VC or a much more reactive comonomer like MMA.

There has been surprisingly little reported on multicomponent polymerization involving VDC considering its industrial importance. Walling and Briggs[67] studied both three and four component reactions involving styrene, MMA, AN and VDC. They obtained excellent agreement between experimental and theoretical compositions using the reactivity ratios reported by Lewis *et al.* Fordyce

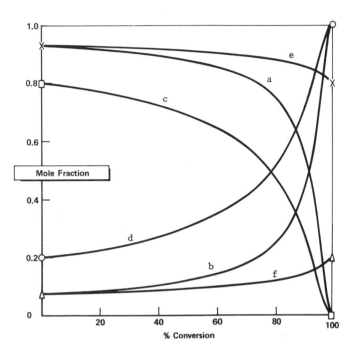

FIGURE 3.4 Theoretical prediction of composition drift in the batch copolymerization of VDC with VC. Initial charge 80 mole % of VDC.
    a: Instantaneous mole fraction VDC in copolymer; b: instantaneous mole
fraction VC in copolymer; c: mole fraction VDC in unreacted monomer;
d: mole fraction VC in unreacted monomer; e: average copolymer composition, mole fraction VDC; f: average copolymer composition, mole fraction
VC.

(Courtesy V. E. Meyer, The Dow Chemical Company, Midland, Michigan.)

et al.[68] showed that *Q-e* values alone gave a satisfactory prediction
of composition. Chan and Meyer[65,69] have analyzed theoretically
a number of systems involving VDC.

3.2.2   Preparative methods of copolymerization

The methods used in batch copolymerization are essentially the same as those already described for the homopolymerization of VDC. The preferred methods are aqueous suspension and emulsion polymerization. Mass reactions are seldom used except to prepare laboratory samples. Homogeneous copolymerizations can be carried out in good solvents because of the greater solubility of the copolymers. But it is difficult to achieve high rates and high molecular weights by this process.

Emulsion and suspension are used commercially for large scale copolymerization.[70] A typical suspension reaction, for a Saran B resin[23] when carried to completion, yields a broad composition distribution copolymer because of the unequal reactivities of VDC and VC. A copolymer of more controlled composition can be prepared by increasing the initial charge of VC and bleeding off VC gas during the run to keep the pressure constant.[71] This tends to keep the unreacted monomer composition fairly constant.

There are two general approaches for making emulsion copolymers depending on the intended use. Low soap recipes of marginal colloidal stability can be used if the polymer is to be isolated and used as a dry powder. When the polymer is to be used as a latex, however, colloidal stability is much more important. These recipes usually include much more surfactant or other colloidal stabilizers.

The major advantage of emulsion copolymerization is again the fact that high molecular weights can be achieved at high reaction rates. Saran B type resins are made in a recipe similar to that described for the homopolymerization of VDC.[72]

Constant composition processes are much easier to design into an emulsion polymerization. The method of continuous monomer addition emulsion polymerization is widely practiced.[73-77] In these systems, a portion of the reactants and emulsifier and most of the water are charged to a reactor to form a seed latex. The remaining ingredients are then metered into the stirred reactor at the desired rates. This allows greater control over both composition distribution and the colloidal properties of the latex.

3.2.3   Block and graft copolymers

The reactions described up to this point yield copolymers with random structures. The preparation of materials with ordered structures such as block and graft copolymers has also been described primarily in the patent literature. These polymers have not been well characterized for the most part and little is known about the chemistry. So only a few examples are included here mainly to illustrate the techniques.

Attempts to graft directly to PVDC are complicated by the insolubility of the polymer in most monomers and by its sensitivity to mechanical, thermal and radiation energy. Grafting to copolymers is much easier. To illustrate, Baer[78] has described a process for grafting PVC to an ethylene/vinylidene chloride copolymer. This was a classical approach in which the copolymer was dissolved in vinyl chloride and the mixture polymerized in suspension with a free radical initiator.

Grafting of PVDC to another polymer substrate is a more common process. For example, an ethylene/vinyl acetate copolymer can be dissolved in VDC and the mixture polymerized free radically to a grafted product.[79,80] In other cases, PVDC has been grafted to the surface of a polymer that does not dissolve in VDC. Gaylord[81] has described a process in which the substrate is irradiated in the present of H_2O_2 with a Van der Graaf accelerator. The solid is then contacted with VDC. Active sites on the surface initiate graft copolymerization.

Two groups have reported more extensive studies of radiation grafting on polymer substrates. Vlasov et al.[20] studied the effect of substrate orientation on the grafting of PVDC to high pressure polyethylene. They irradiated the substrate in the presence of VDC vapor with x-rays, at 6 rad/sec, 60 °C and 500 mmHg pressure. PVDC was obtained at the rate of 5% per hour with a yield of 15 monomer units converted per electron volt. The PVDC deposited in an oriented structure indicating solid state growth.

Kawase and Hayakawa[82] presorbed the monomer onto polypropylene (PP) fibers and then irradiated with γ-rays. They measured the amount of polymer formed gravimetrically during irradiation. They found the amount grafted dropped off with increasing temperature as shown in Figure 3.5. The amount of grafting increased with initial vapor pressure of VDC and with dose rate. Whether or not all the PVDC formed was grafted to the PP

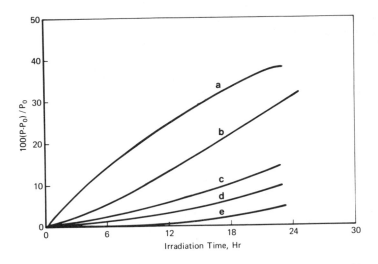

FIGURE 3.5 Grafting of vinylidene chloride at various temperatures: (a) 0 °C; (b) 25 °C; (c) 39 °C; (d) 50 °C; (e) 70 °C. Vinylidene chloride, 210 mmHg; dose rate, 1.2×10^4 rad/h (Ref. 82).

was not clearly established.

Several free radical techniques for making block copolymers have been suggested. Hiemeleers and Smets[83] made polyacrylonitrile by UV irradiation to get a high concentration of trapped radicals at the chain ends. Contacting the polymer with VDC and heating yielded only a small amount of PVDC in the block, < 2%.

Ceresa[84] reported a somewhat more successful approach. He copolymerized MMA and oxygen to a polymer with main chain peroxy links. Heating this polymer in the presence of VDC yielded a block copolymer. The analysis indicated 57% PMMA and 43% of a MMA/VDC block containing 29.6% VDC.

Most 'living polymer' approaches are not suitable for making block copolymers with VDC. While it does polymerize anionically, the polymer also reacts with the living chain ends. Recently, a method which supposedly places short VDC blocks on the ends of polyolefin chains has been described.[85] The olefin is first polymerized with certain Ziegler catalysts. The reaction is partially completed before adding VDC. The final products contain very low

VDC contents, (1-2%), but no evidence is presented that these are true blocks.

3.2.4   Ionic copolymerization

The best evidence for an ionic mechanism copolymerization comes from a comparison of reactivity ratios. Konishi[86] has done this to establish the anionic character of the polymerization of VDC initiated by n-butyl lithium. His results shown in Table 3.6 clearly demonstrate that the process is not a free radical reaction.

TABLE 3.6
Comparison of Reactivity Ratios

Monomer 2	Free radical		Anionic	
	r_1	r_2	r_1	r_2
Methylacrylate	1.0	1.0	0.005	38
Styrene	0.14	2.0	0.0015	3.4
Vinylchloride	3.25	0.3	0.001	0.05

Yamazaki et al.[27] used a Ziegler type catalyst to copolymerize VDC and VC. The system was $Ti(OC_4H_9)_4$-$Al(C_2H_5)Cl_2$ in hexane at 30 °C. The reactivity ratios were (VDC = Monomer 2) r_1 = 0.59 r_2 = 0.90. These do not agree with Konishi's values suggesting that the process is not a simple anionic polymerization. The British patent cited earlier[28] also described the preparation of various VDC copolymers. Among these were VC, propylene, VAC, MMA and MA. The data reported for the VDC/VC system indicate that VDC is slightly less reactive than VC (Table 3.7).

The low temperature radiation copolymerization of VDC and IB has been studied by Sheinker et al.[15] They observed a dramatic change in reactivity ratios for the temperature range −78 to 0 °C. At the highest temperature, both copolymer compositions and rates indicated a free radical reaction. At −78 °C, the reaction was characterized as cationic. It is known that IB polymerizes to high molecular weight easily by a cationic mechanism while VDC does not; the opposite is true for free radical reactions. The reactivity ratios they calculated are shown in Table 3.8. The data in Table 3.8

TABLE 3.7

Effect of Feed Composition on Polymer Composition for the
Copolymerization of VDC and VC with a Ziegler Type Catalyst

Example no.	VDC	VC[a]	Polymer yield %	VC content in copolymer (mole %)
118	7.0	1.0	22.5	17.6
119	6.0	2.0	22.9	20.9
120	5.0	3.1	6.4	37.6
121	4.0	4.2	7.2	27.2
122	3.0	5.0	2.0	53.8
123	2.0	6.0	5.5	74.4

[a]   VC: volume at $-78\ ^\circ$C.

TABLE 3.8

Effect of Temperature on the γ-ray initiated Copolymerization
of VDC and IB (r_1 — isobutylene, r_2 — vinylidene chloride)

Polymerization temperature, $^\circ$C	With 20% ZnO		Without ZnO	
	r_1	r_2	r_1	r_2
− 79.5	56	0.02	25	0
− 39	1.5	0.11	1.27	0.21
0	0.40	1.00	0.03	1.3

also show that the copolymerization of VDC and IB is influenced
significantly by the presence of ZnO.[16] Again, this is more likely to
occur in an ionic mechanism. Reactivity ratios in the copolymerization of VDC and IB effected by chemical free radical initiators are
not very temperature sensitive,[51] indicating that the ionic reaction
is radiation induced.

    Other attempts to induce cationic copolymerization of VDC
have been unsuccessful.[14,87,88]

REFERENCES

1. P. J. Flory, *Principles of Polymer Chemistry*, Cornell Univ. Press, Ithaca, N.Y., Chap. IV (1953).
2. G. Talamini and E. Peggion in G. E. Ham (ed.), *Vinyl Polymerization*, Vol. 1, Part I, M. Dekker, N.Y., Chapter 5 (1967).
3. C. H. Bamford, W. G. Barb, A. D. Jenkins and P. F. Onyon, *The Kinetics of Vinyl Polymerization by Radical Mechanisms*, Butterworth, London, Chapter 4 (1958).
4. C. E. Schildknecht, *Vinyl and Related Polymers*, Wiley and Sons, N.Y., Chapter VIII (1952).
5. J. J. P. Staudinger, *British Plastics*, **20**, 381 (1947).
6. R. C. Reinhardt, *Ind. Eng. Chem.*, **35**, 422 (1943).
7. R. M. Wiley, The Dow Chemical Company, private communication.
8. C. E. Bawn, T. P. Hobin and W. J. McGarry, *J. Chim. Phys.*, **56**, 791 (1959).
9. W. A. Strong, *Ind. Eng. Chem. Prod. R & D*, **3**, 264 (1964).
10. G. Talamini and G. Vidotto, *Chim e Ind.*, **46**, 371 (1964).
11. J. D. Burnett and H. W. Melville, *Trans. Faraday Soc.*, **46**, 976 (1950).
12. C. H. Bamford and A. D. Jenkins, *Proc. R. Soc., Lond.*, **A216**, 515 (1953).
13. R. F. Boyer, *J. Phys. Colloid Chem.*, **51**, 80 (1947).
14. W. J. Burlant and D. H. Green, *J. Polym. Sci.*, **31**, 227 (1958).
15. A. P. Sheinker *et al.*, *Dokl. Akad. Nauk SSSR*, **124**, 632 (1959).
16. E. V. Kristal'nyi and S. S. Medvedev, *Vysokomol. soyed.*, **7**, 1373 (1965).
17. G. Ley, C. Schneider and D. Hummel, *J. Polym. Sci.*, **C27**, 119 (1969).
18. J. F. Brown and D. M. White, *J. Am. Chem. Soc.*, **82**, 5671 (1960).
19. M. A. Bruk and V. I. Lukhovitskii, *Polym. Sci. USSR*, **5**, 1004 (1966).
20. A. V. Vlasov *et al.*, *Dokl. Akad. Nauk SSSR*, **161**, 857 (1965).
21. H. Wiener, *J. Polym. Sci.*, **7**, 1 (1951).
22. P. M. Hay, J. C. Light, L. Marker, R. W. Murray, A. T, Santonicola, O. J. Sweeting and J. G. Wepsic, *J. Appl. Polym. Sci.*, **5**, 23 (1961).
23. L. C. Friedrich, Jr., T. W. Peters and M. R. Rector, U.S. 2,968,651, to The Dow Chemical Company (1961).
24. E. N. Zilberman, A. E. Kulikova, N. M. Pinchuk, N. K. Taikova and N. A. Okladnov, *J. Polym. Sci. A-1*, **8**, 2325 (1970).
25. A. Konishi, *Bull. Chem. Soc. Japan*, **35**, 197 (1962).
26. B. L. Erusalimskii *et al.*, *Dokl. Akad. Nauk. SSSR*, **169**, 114 (1966).
27. N. Yamazaki, K. Sasaki, T. Nisiimura and S. Kambara, *ACS Polymer Preprints*, **5**, 667 (1964).
28. Brit. 1, 119, 746, to the Chisso Corp., Japan (1967).
29. T. Alfrey, Jr., J. J. Bohrer and H. Mark, *Copolymerization*, Interscience, N.Y. (1952).
30. G. E. Ham (ed.), *Copolymerization*, Interscience, N.Y. (1964).

31.   F. M. Lewis, F. R. Mayo and W. F. Hulse, *J. Am. Chem. Soc.*, 67, 1701 (1945).
32.   T. Alfrey, Jr. and C. C. Price, *J. Polym. Sci.*, 2, 101 (1947).
33.   J. P Kinsinger, T. Fischer and C. W. Wilson, III, *Polym. Lett.*, 5, 285 (1967).
34.   E. F. Jordan, Jr., K. M. Doughty and W. S. Port, *J. Appl. Polym. Sci.*, 4, 203 (1960).
35.   H. Mark, B. Immergut, E. H. Immergut, L. J. Young and K. I. Kenyon, in J. Brandrup and E. H. Immergut (eds.), *Polymer Handbook*, Interscience, N.Y. (1966), pp. II-265 to II-268. L. J. Young, *ibid.*, pp. II-333 to II-334.
36.   E. H. Hill and J. R. Caldwell, *J. Polym. Sci.*, 47, 397 (1960).
37.   R. B. Parker and B. V. Mokler, *Polym. Lett.*, 2, 19 (1964).
38.   L. Marker, O. J. Sweeting and J. G. Wepsic, *J. Polym. Sci.*, 57, 855 (1962).
39.   M. Ichida and H. Nagao, *Bull Chem. Soc., Japan*, 30, 314 (1957).
40.   C. S. Marvel, T. K. Dykstra and F. C. Magne, *J. Polym. Sci.*, 62, 369 (1962).
41.   A. Winston and G. T. C. Li, *J. Polym. Sci.*, A-1, 5, 1223 (1967).
42.   U. Johnsen, *Kolloid-Z. Z. Polym.*, 210, 1 (1966); *Ber. Bunsenges. Phys. Chem.*, 70, 320 (1966).
43.   V. E. Meyer, *J. Polym. Sci.*, A4, 2819 (1966); *ibid.*, 5, 1289 (1967).
44.   P. Agron, T. Alfrey, Jr., J. Bohrer, H. Haas and H. Wechsler, *J. Polym. Sci.*, 3, 157 (1948).
45.   J. G. Cobler, M. W. Long and E. G. Owens, in O. J. Sweeting (ed.), *The Science and Technology of Polymer Films*, Vol. I., Interscience, N.Y., Chapter 15 (1968).
46.   J. L. McClanahan and S. A. Previtera, *J. Polym. Sci.*, A3, 3919 (1965).
47.   R. Chujo, S. Satoh and E, Nagai, *J. Polym. Sci.*, A2, 895 (1964).
48.   R. Chujo, S. Satoh, T. Ozeki and E. Nagai, *J. Polym. Sci.*, 61, S12 (1962).
49.   T. Fischer, J. B. Kinsinger and C. W. Wilson, *Polym. Lett.* 4, 379 (1966).
50.   K. H. Hellwege, U. Johnsen and K. Kolbe, *Kolloid-Z. Z. Polym.*, 214, 45 (1966).
51.   U. Johnsen and K. Kolbe, *Kolloid-Z. Z. Polym.*, 232, 712 (1969).
52.   U. Johnsen and K. Kolbe, *Kolloid-Z. Z. Polym.*, 216-217, 97 (1967).
53.   K. Ito, S. Iwase and Y. Yamashita, *Makromol. Chem.*, 110, 233 (1967).
54.   Y. Yamashita, K. Ito, S. Ikuma and H. Kada, *Polym. Lett.*, 6, 219 (1968).
55.   K. Ito and Y. Yamashita, *Polym. Lett.*, 6, 227 (1968).
56.   U. Johnsen and W. Lesch, *Kolloid-Z. Z. Polym.*, 233, 863 (1969).
57.   H. Germar, *Makromol. Chem.*, 84, 36 (1965).
58.   S. Krimm and C. Y. Liang, *J. Polym. Sci.*, 22, 95 (1956).
59.   S. Narita, S. Ichinohe and S. Enomoto, *J. Polym. Sci.*, 36, 389 (1959).
60.   S. Enomoto, *J. Polym. Sci.*, 55, 95 (1961).

61.   M. R. Meeks and J. L. Koenig, *J. Polym. Sci.*, A-2, 9, 717 (1971).
62.   S. Tauge, T. Okumoto and T. Takeuchi, *Makromol. Chem.*, 123, 123 (1969).
63.   O. L. Stafford, Paper presented at Fall Meeting, ACS, Midland Section (1970).
64.   V. E. Meyer and G. G. Lowry, *J. Polym. Sci.*, A3, 2843 (1965).
65.   R. K. S. Chan and V. E. Meyer, *J. Polym. Sci. C*, 25, 11 (1968).
66.   G. E. Molau, *J. Polym. Sci.*, A5, 401 (1967).
67.   Walling, C. and E. R. Briggs, *J. Am. Chem. Soc.*, 67, 1774 (1945).
68.   R. G. Fordyce, E. C. Chaplin and G. E. Ham, *J. Am. Chem. Soc.*, 70, 2489 (1948).
69.   R. K. S. Chan and V. E. Meyer, *J. Makromol. Sci.*, A1, 1089 (1967).
70.   J. F. Gabbett and W. Mayo Smith, in *Copolymerization*, Interscience, N.Y., Chapter 10 (1964).
71.   J. Heerema, U.S. 2,482,771, to The Dow Chemical Company (1944).
72.   P. K. Isaacs and A. Trofimow, U.S. 3,033,812, to W. R. Grace (1968).
73.   D. S. Gibbs and R. A. Wessling, U.S. 3,617,368, to the Dow Chemical Company (1971).
74.   P. K. Isaacs, D. G. Woodward, A. Trofimow and D. M. Wacome, U.S. 3,317,449, to W. R. Grace (1967).
75.   W. G. MacPherson and C. D. Parker, U.S. 2,956,047, to The Dow Chemical Company (1960).
76.   D. M. Woodford, U.S. 3,291,761, to Scott Bader (1967).
77.   D. M. Woodford, *Chem. Ind.*, 316 (1966).
78.   M. Baer, U.S. 3,366,709, to Monsanto (1968).
79.   R. Buning and W. Pungs, Can. 798,905, to Dynamit Nobel (1968).
80.   H. Bartl and D. Hardt, *Adv. Chem. Ser.*, 91, 477 (1969).
81.   N. G. Gaylord, U.S. 2,907,675, to DuPont (1959).
82.   K. Kawase and K. Hayakawa, *J. Polym. Sci.*, A-1, 7, 3363 (1967).
83.   J. Hiemeleers and G. Smets, *Makromol. Chem.*, 47, 7 (1961).
84.   R. J. Ceresa, *Polymer*, 1, 397 (1960).
85.   H. J. Hagemeyer and M. B. Edwards, U.S. 3,453,346, to Eastman (1969).
86.   A. Konishi, *Bull. Chem. Soc. Japan*, 35, 395 (1962).
87.   C. E. R, Walling, Briggs, W. Cummings and F. R. Mayo, *J. Am. Chem. Soc.*, 72, 48 (1950).
88.   R. E. Florin, *J. Am. Chem. Soc.*, 71, 1867 (1949).

CHAPTER 4

Heterogeneous Polymerization

4.1 SLURRY REACTIONS

The classical theories of free radical polymerization assume homo-
geneity of the reaction medium. Most theories also include the
assumption of constant reaction volume. Because of these limita-
tions, such theories can be applied to vinylidene chloride only in
the relatively uninteresting case of a batch solution polymerization
in a good solvent. Significant modifications are required to make
them applicable to the cases of greatest interest: slurry, suspension
and emulsion polymerization.

The heterogeneous nature of the mass polymerization of VDC
was described in Chapter 3. A typical conversion-time curve was
shown in Figure 3.1 for a free radical mass polymerization. The
reaction can be divided into three stages: the first stage during which
the rate is rapidly increasing; the second stage in which the rate has
a fairly constant value, and the third stage in which the rate shows a
gradual decrease to zero due to depletion of monomer supply.
Since the reaction mixture solidifies while 70-80% of the monomer
is still unreacted, further polymerization generates void space. The
product at high conversion is a hard, porous, opaque solid.

A number of common monomers, including vinyl chloride and
acrylonitrile also polymerize heterogeneously. The process is complicated and a completely satisfactory mechanism for these reactions
has not been developed.

Jenkins[1] has recently reviewed the area of heterogeneous polymerization. Experimental results suggest that the mechanism
involves a two-phase reaction. The liquid phase reaction follows
normal kinetics, but with the possibility also of transferring

radicals to the solid polymer phase. The latter is present because of the low solubility of the polymer in the liquid phase. The novel feature of this mechanism is the proposal that polymerization can take place also in the solid phase but the rate of termination is reduced. The autoacceleration observed in these systems has been attributed to the latter phenomenon. The data indicate that the solid polymer phase plays a key role. No abnormalities are observed in the kinetics when these monomers are polymerized in a homogeneous system.

There have been many attempts to derive rate equations for heterogeneous polymerization, but the role of polymer morphology was not specifically considered in any of the older work[2-8]. Recent papers by Mazurek[9] and Cotman et al.[10] on the kinetics of VC polymerization, Lewis and King[11] on acrylonitrile and Wessling and Harrison[12,13] on VDC, have identified the importance of the polymer properties and morphology.

4.1.1   Experimental results on VDC

The extension of theories derived for VC or AN to VDC must be considered with caution. The similarity between these cases may be only superficial. The plots of conversion versus time follow the same characteristic pattern; but one cannot draw any conclusions from this fact alone since many other aspects of VDC polymerization are significantly different.

Burnett and Melville[14] carried out an elaborate series of experiments designed to measure the rate constants for the polymerization of VDC. They followed the course of a photoinitiated reaction dilatometrically and used the techniques of intermittent illumination and retardation by benzoquinone to measure radical lifetimes and rate of initiation. The rate constants, k_p and k_t, were calculated from these data.

This study has been criticized on several counts. According to Arlman and Wagner,[15] the factor used by Burnett and Melville to convert volume change to fraction polymerized was seriously in error. Based on the densities of liquid monomer and pure crystalline polymer, the ratio,

$$\frac{\Delta V}{\Delta C} = 0.385$$

Arlman and Wagner obtained an experimental value of 0.39 compared to 0.26, the value used by Burnett and Melville. Therefore, the rates reported by the latter are too large by a factor of approximately 1.5.

When the only set of experimental data (ΔV vs t) included in this paper are replotted using the correct factor, the curvature at short times is unmistakeable. An increasing rate in the early stages is characteristic of heterogeneous polymerizations so its appearance in this case is not surprising. Burnett and Melville calculated the rate from the average slope. The value they reported (0.10% per minute) is significantly higher than the initial slope (0.056% per minute) of the corrected plot.

The drop-off in rate in photoinitiated heterogeneous reactions has been attributed by Bamford and Jenkins[16] to a reduction in light intensity caused by scattering in the increasingly turbid reaction mixture. They suggest that the apparent linearity is a result from the increasing polymerization rate and decreasing rate of initiation with increasing conversion. Thus, photoinitiation techniques are not suitable for kinetic studies of heterogeneous reactions.

The kinetic analysis made by Burnett and Melville can only have relative significance because of the above described problems. However, the dependence on rate of $I^{1/2}$ where I is the intensity of incident illumination seems quite well established by their studies. The first order dependence of rate on monomer concentration in cyclohexane should also be valid.

Burnett and Melville calculated the activation energies for the overall process from kinetic studies at 15, 25 and 35 °C. They found a slight dependence of ΔE^* on light intensity but the average value 5.0 ± 0.2 kcal/mole is in the range normally observed for photoinitiated polymerizations. The intrinsic viscosity of the product increased with polymerization temperature indicating that the rate of initiation was temperature independent.

Abnormal behavior was observed only in the temperature dependence of the derived rate constants. This is of doubtful significance, but has been interpreted to mean that radicals become trapped in the precipitating polymer. This produces an abnormally low but very temperature dependent termination constant. While such an interpretation may be correct, it cannot be substantiated by these data.

Bengough and Norrish[17] used a chemical initiator, benzoyl peroxide, to study the mass polymerizations of VDC. The conversion-time curves they observed are shown in Figure 4.1. In a series of experiments, they were able to demonstrate conclusively that the increase in rate was not due to impurities but was associated with the solid polymer phase. Conversion-time plots in tetrahydrofuran (THF) at 61 °C were linear in the low conversion regions. THF swells and partially dissolves PVDC at this temperature. Another series of experiments clearly demonstrated the presence of surface radicals. They stopped the polymerization at low conversion by distilling off the remaining monomer. If the solid polymer was exposed briefly to oxygen, and then reimmersed in the monomer, the polymerization started slowly as though no solid were present. If the polymer were not exposed, however, the monomer when added back began to polymerize at the accelerated rate existing at the point of interruption.

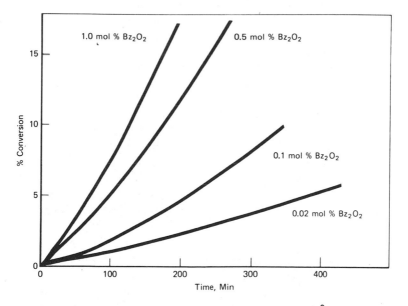

FIGURE 4.1 The polymerization of vinylidene chloride at 47 °C with various concentrations of benzoyl peroxide (Ref. 17).

Bengough and Norrish found that the rapidly increasing rate at low conversion made it difficult to obtain the initial rate. But rates obtained from the slopes of the conversion-time curves at 5% conversion were proportional to the half power of the Bz_2O_2 concentration. They made no attempt to derive rate constants but observed that the theoretical rate equation they had used earlier to describe the polymerization of VC fit these data also suggesting a similar mechanism.

The apparent activation energy of the reaction in the range of 47-75 °C had the unusually low value of 15.4 kcal/mole. Since the accepted value for the ΔE^* of the benzoyl peroxide decomposition is 30 kcal/mole, this result implies that the activation energy of polymerization, $(E_p - \frac{1}{2}E_t)$ is approx 0.4 kcal/mole. The poor agreement with the value obtained by Burnett and Melville suggests that the activation energies may be temperature dependent, but there is also the possibility that the 3.4 value is in error. The smaller value (0.4) would require that E_p and E_t are of the same magnitude. More data are needed to resolve this point.

While both of the above described studies suggest a solid phase reaction, other observations tend to cloud the issue. A long photo after effect was not observed as might be expected, and Bamford et al.[18] could detect no radicals in PVDC polymerized at 20 °C. Bawn et al.[19] were able to generate trapped radicals by polymerizing with γ-rays at temperatures from 25-80 °C. They found a very low post-irradiation rate and could only induce a slight 'fast' reaction by heating. They concluded that the radicals were present in low concentration and could react only if the crystalline phase was partially destroyed. The study did not distinguish between radical terminated chains buried in the crystal and radicals formed by radiation damage to the polymer. PVDC is known to degrade on exposure to γ-rays.[20] It appears that while evidence for trapped radicals in PVC and PAN is strong, the situation with PVDC is not so clear cut.

In comparing PVC or PAN to PVDC, we should keep in mind the very signficant differences in the solid state structures of these materials. First of all, PVDC is semi-crystalline and is well above its glass transition temperature over the range of temperatures in which kinetics have been studied. The vinyl monomers, on the other hand, are converted to glassy solids. (Both PVC and PAN are slightly crystalline, however.) PVDC is highly crystalline because it has a linear symmetrical chain structure which is independent of the temperature of polymerization. Talamini and Vidotto[21] in a study of the effect of polymerization temperature on molecular weight found no abnormalities. They note that the crystallinity of these polymers is virtually independent of polymerization temperature. The molecular weight falls with increasing temperature as would be expected for chemical initiation.

The structures of PVC and PAN are affected by reaction temperature. This changes properties such as crystallinity and solubility. The solubility of PVDC is only mildly affected by temperature of polymerization.[22] The observed effects must be due to changes in morphology only since the structure of the polymer is unaffected. Differences in solubility except those due to molecular weight can be erased by conditioning the samples with an identical thermal cycle, e.g., first melting, then recrystallizing at a fixed temperature.

The solubility of PVDC in its monomer or in common solvents in the temperature range normally used for polymerization studies (0-60 °C) is very low. The particles remain essentially unswollen and undergo no change in properties. Therefore, if the solid phase contains no monomer except that adsorbed on its surface, the interior of the crystals should remain inaccessible unless the polymer is heated to a high enough temperature for chain motion in the crystalline phase to occur. In the solid state, motion in the crystalline phase is observed above approximately 80 °C.[23] This temperature could be somewhat lower depending on the medium in which the solid was suspended.

The dilatometric studies of Arlman and Wagner mentioned earlier suggest that 'as polymerized' PVDC is nearly 100% crystalline. Several studies in more recent times have shown that during polymerization, it precipitates in the form of reasonably well-defined lamellar crystals.[13,24] The morphology of these crystals depends on the polymerization conditions, particularly on the solvent medium. Marginal solvents[22] (those that dissolve PVDC at approximately 100 °C) yield polymer crystals that are very similar to those grown by quenching a dilute polymer solution. The crystal habits developed by polymerization in a non-solvent, on the other hand, have not been duplicated by solution crystallization.[13] (Morphology of PVDC is discussed in Chapter 6.)

4.1.2   Mechanism

The morphological studies of "as polymerized" PVDC[23] lead to the conclusion that there should be two classes of polymerization reactions: one in which the polymer forms in solution and then precipitates; the other in which polymerization takes place on the surface of the solid phase. However, the kinetic data fit a theoretical model for the kinetics that assumes a solid state growth.[12] The elementary reactions in this mechanism are listed below:

TABLE 4.2

Elementary Reactions in the Mechanism of Heterogeneous
Free Radical Polymerization (Ref. 12).

Step 1.   Initiation in the liquid phase

$$I \xrightarrow{k_i} 2\,R\cdot$$

Step 2.   Propagation in the liquid phase

$$R\cdot + M \xrightarrow{k_p} P\cdot$$

Step 3.   Termination in the liquid phase

$$2\,P\cdot \xrightarrow{k_t} P$$

Step 4.   Radical precipitation

$$P\cdot \xrightarrow{k_c} P\cdot_s$$

Step 5.   Propagation at solid-liquid interface

$$P\cdot_s + M \xrightarrow{k_p^s} P\cdot_s$$

Step 6.   Termination at the solid-liquid interface

$$2\,P\cdot_s \xrightarrow{k_t^s} P$$

The initial rate for this mechanism is the same as that for a homogeneous reaction:

$$\frac{1}{M_0}\frac{dm}{dt} = R_p = \left(\frac{f k_i k_p^2}{k_t}\right)^{1/2} [I]_0^{1/2} (1 - m/M_0), \quad (4.1)$$

where M_0 is the moles of monomer present initially, m is the moles converted and $[I]$ is the initiator concentration.

At some point early in the reaction the solution will become supersaturated with polymer and polymer crystals will begin to precipitate. Under normal conditions, such as mass polymerization at ambient temperatures, the solid phase appears at very low conversions. From this point on, the effect of the solid phase must be considered.

Two limiting cases of interest are derivable from the general analysis:

Case I   Solution polymerization followed by precipitation

The rate equation for mass polymerization is:

$$R_p = R_p^0(1 - m/M_0)^{1/2}, \quad (4.2)$$

where R_p^0 is the same as the initial rate in a homogeneous reaction. In this case, the solid phase has no direct effect on the kinetics.

Case II   Surface polymerization

The basic assumptions here are that polymerization occurs only on the solid phase surface and that radicals precipitate before terminating in the liquid phase. This means that Step 3 can be eliminated from the mechanism.

In this limiting case the rate is dependent on the amount of solid phase; the rate increases with conversion because of the increase in surface area. The relationship of surface area to the amount of polymer present is determined by the number of particles, their shape, and the loci of polymerization.

Neither particle number nor shape are well defined in mass polymerizations, but morphological studies have suggested as a model, a rectangular lamellar particle growing on the edges only. The rate equation for this case is

$$R_p = K C_1 [I]_0^{1/2} (N/M_0)^{1/4} (m/M_0)^{1/4}, \qquad (4.3)$$

where K is a kinetic factor:

$$K = \left[\frac{f k_i k_p s^2}{k_t s} \right]^{1/2}$$

and C_1 is a morphology factor:

$$C_1 = \left[\frac{2 d (q + 1)}{V_M} \quad \frac{h V_p}{q}^{1/2} \right]^{1/2},$$

where q is the ratio of large lamellar dimensions, h is the fold length, d the thickness of the reaction zone, and V_M the molar volume.

This analysis predicts that a plot of $(m/M_0)^{3/4}$ vs t should be linear. This turns out to be the case for polymerization both in mass and in dioxane. The former was anticipated but in view of the "as polymerized" morphology of PVDC prepared in dioxane, a case I reaction would be expected. However, the rate appears to increase with conversion as in a mass reaction.

4.1.3   Copolymerization in slurry reactions

Mass copolymerizations yielding high VDC content copolymers are also normally heterogeneous. Two of the most important pairs, VDC/VC and VDC/AN, are heterogeneous over the entire composition range. In both cases, at either composition extreme, the product separates initially in a powdery form, but for intermediate compositions the reaction mixture may only gel. Copolymers in this composition range are swollen but not dissolved completely by the monomer mixture at normal polymerization temperatures. Copolymers containing more than 10-15 mole % acrylate are normally soluble. These reactions are, therefore, homogeneous, and if carried to completion, yield clear solid castings of the copolymers. In cases where the composition drift is severe, the reaction may be homogeneous during one stage of the reaction and heterogeneous in another. For example, we would expect high VDC content co-polymer to precipitate near the end in a copolymerization with MMA. The opposite would occur if the comonomer were vinyl acetate. It should be expected that in any case where phase separation occurs, both kinetics and copolymer composition can be affected.

Rates of copolymerization of VDC with small amounts of a second monomer are sometimes lower than its rate of homopoly-merization. The dependence of rate on monomer composition for the VDC/VC system is shown in Figure 4.2.[25] The kinetics of this system were analyzed by Melville[26] who estimated that the high rate of cross termination could account for the decrease in rate with increasing vinyl chloride concentration. Lower rates of copolymerization have also been reported for copolymerization of vinyl acetate and styrene with VDC. The latter, in low concentrations, acts almost as an inhibitor.

The kinetics of the copolymerization of VDC and VC were studied by Bengough and Norrish.[25] As was true of the respective homopolymerizations, the rates of copolymerization increased with conversion over the entire composition range. The curve calculated from the measured reactivity ratios fit the data quite well, suggesting that copolymerization theory does apply. However, the fit may be fortuitous since the character of the solid phase varies from highly crystalline to swollen gel to hard glassy solid as the concentration of vinyl chloride in the polymerization mixture varies from 0 to 100%.

4.2   KINETICS OF POLYMERIZATION IN AQUEOUS MEDIA

The two basic processes used to polymerize a water insoluble monomer in aqueous media are suspension and emulsion poly-merization. Kinetically, they are very distinct processes. Suspen-sion reactions use oil soluble initiator dissolved in the monomer. The solution is dispersed in water before reaction. Each monomer droplet is like a small mass polymerization. The kinetics of this process are similar to that of a mass reaction. The rate increases with conversion in the early stages; during the course

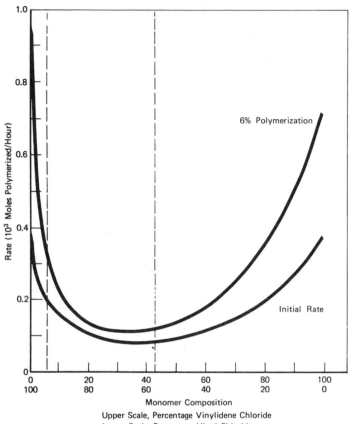

FIGURE 4.2 The variation of the rate of copolymerization with monomer concentrations at 47 °C. Polymer gel was observed in the region between the dashed lines (Ref. 25).

of the reaction, the particles solidify. Since each monomer droplet eventually becomes a polymer particle, the size of the latter is determined by the degree of dispersion of the monomer suspension. Sizes in the range of 25-100 μ or more are commonly observed.

The rates of suspension polymerization may be somewhat higher than in mass due to the subdivision of the reaction into small droplets. But this aspect has not been studied quantitatively. In any case, the rates do not reach the levels attainable in emulsion.

Water soluble initiators and soaps are employed in emulsion polymerization. The initial loci of polymerization are believed to be the soap micelles which grow into soap stabilized polymer particles. Substantial polymerization does not occur in the monomer droplets. Hence, the particle size is not dependent on the degree of monomer dispersion but rather on the soap level. Particle sizes in the range of 500-3000 Å are typically obtained.

The number of particles is fixed early in the reaction and further polymerization takes place either on or in the latex particle. Normally radicals are generated in the aqueous phase and monomer is added as a third phase. Both reactants, therefore, must diffuse through the aqueous phase to reach the reaction site. Some small amount of polymerization may occur in the aqueous phase, but the major part must be associated with the polymer particles.

There are two basic models used to describe the reaction site: the swollen particle model first analyzed by Smith and Ewart[27] and the surface growth model.[28] In the swollen particle model, the latex particle is swollen uniformly with monomer and the entire volume of the particle is available for reaction. Since the reactants diffuse in through the surface, the model contains the implicit assumption that diffusion throughout the particle is not rate determining. This implies both low particle viscosity and small particle size. Since polymer is being formed everywhere within the particle, it must of necessity grow uniformly.

The surface model assumes that the polymerization reaction takes place in a restricted zone at the surface. This could be an adsorbed monomer layer, or a highly swollen surface. Such a particle would, as a consequence, grow from the surface outward. The surface model requires that no reaction takes place in the core of the particle. However, it does not require that the particles be completely unswollen in the interior.

4.2.1 Batch emulsion polymerization[27,29,30]

Typically a batch emulsion polymerization can be subdivided into three stages: particle formation, particle growth, and finishing. The Smith Ewart analysis predicts that the number of particles, N, formed in the first stage depends on the soap and initiator levels

$$N \alpha \, [I]^{2/5} \, [S]^{3/5} , \tag{4.4}$$

where $[I]$ is the initiator concentration and $[S]$ is the soap concentration. The rate equation for the growth stage using the swollen particle model is

$$R_p = \frac{k_p \, N \, [M] \, Q}{N_A} , \tag{4.5}$$

where R_p = dm/dt (m = moles of monomer converted)
N = number of particles
N_A = Avogadro's Number
$[M]$ = concentration of monomer in the particles (moles/liter)
Q = number of radicals per particle.

In the limiting case of very small particle size, $Q = 1/2$ and $[M]$ is approximately constant. Hence, this theory predicts that the rate is directly proportional to N in the growth stage and is constant when N is constant, the case normally encountered. For the latter, conversion in Stage II increases linearly with time:

$$m = m_0 + R_p * t , \tag{4.6}$$

where R_p is the Smith-Ewart limiting rate; and m_0 is the moles of monomer converted in the seed reaction.

Gardon[30] has treated in detail the case in which Q increases with particle size. He obtains a time dependent rate.

$$R_p = R_p * + 2 A t , \tag{4.7}$$

where the constant, A, is independent of the number of particles. This factor compensates for the reduced rate of termination in larger size particles. In the Gardon analysis, m is a quadratic function of time.

In order to formulate a mathematical model for the surface reaction, it can be assumed that a steady state radical concentration is generated on the particle surface.[28,31] This leads to the result that the rate should be insensitive to the number of particles; but that may be only a consequence of that assumption.

The governing equation for a surface reaction model is essentially of the same form as Equation (4.5).

$$R_p = \frac{k_{p_s} N}{N_A} \frac{M_s Q_s}{V_s},$$ (4.8)

where the subscripts denote surface quantities rather than the bulk quantities of Equation (4.1). One conclusion of this model is that if the particles are small enough, Q_s will also have a value of $1/2$. In the presence of excess monomer, M_s/V_s will be approximately constant and, therefore, the rate equation will of the same form regardless of whether a surface or swollen particle reaction is postulated. This suggests that kinetic studies cannot be used to distinguish between these mechanisms.

The other limiting case is that of a very large particle having a steady state radical concentration on its surface. The rate for this model depends on the surface area of the polymer phase and increases with conversion. The steady state assumption leads to the prediction that the number of radicals per particle is proportional to the square root of the surface area. This also contains the assumption that the thickness of the reaction zone is constant. The rate equation for this case is:

$$R_p = KN^{1/6}[m_0^{2/3} + 2/3\, KN^{1/6}t]^{1/2}.$$ (4.9)

The conversion is also a non-linear function of time, but the surface area increases linearly with time.

4.2.2   Experimental results, emulsion polymerization

The first studies of the emulsion polymerization of VDC were aimed at developing processes for making Saran resins and little of this work examined fundamental kinetic aspects. Later, extensive kinetic studies were undertaken by Moll and coworkers at The Dow Chemical Company. While these studies have never been published, some of the conclusions drawn from them were discussed in the Styrene Monograph.[32] Among other studies carried out in this period, Staudinger[33] showed conversion/time curves for emulsion polymerization using both oil soluble and water soluble initiators. Much faster rates were obtained with the latter. The rate was found to increase with the initiator concentration in either case.

In 1951, Wiener[34] described the persulfate-initiated polymerization of VDC at 25 °C, both in soap solution and in emulsion. He used potassium laurate as the emulsifier and followed the kinetics dilatometrically at various soap levels. The conversion/time curves were linear up to 60% conversion. His results were in agreement with the Smith-Ewart theory.

Tkachenko and Khomikovskii[35] studied the emulsion polymerization of VDC in a sulfonate soap system at low pH. They kept the reaction saturated with monomer vapor. In the range of 3-10% soap, at 40 °C, the conversion-time curves were not linear. In fact, the rate started at a high level, fell to zero and then climbed again to a somewhat lower value. They attributed the initial rate to reaction in micelles to the point where solubilized monomer was used up. Further reaction was attributed to particle growth with polymerization on the surface. They suggested that the constant rate was caused by constant surface area of the polymer phase, but gave no evidence for this hypothesis. The activation energy in the growth stage was given as 14 kcal/mole.

Hay and coworkers[36-38] also observed the break in the conversion-time curve. Their results were obtained using a redox initiator system, $(NH_4)_2 S_2 O_8 / Na_2 S_2 O_5$, with sodium lauryl sulfate as emulsifier. They carried out an intensive study of the kinetics in this system hoping to find an explanation for the sudden drop in rate.

The conversion-time curves were sensitive to initiator level, soap concentration and rate of stirring. At low rates, or with rapid stirring, conversion increased smoothly with time. The break was most pronounced in reactions carried out at a high initial rate with slow stirring.

· The reaction was considered as a three stage process as shown in Figure 4.3. The rate was proportional to t^2 in the seed step. The dependence of particle concentration on initiator and soap concentration followed the form predicted by the Smith-Ewart theory.

The drop off in rate (Stage 2) occurred at a point where the seed reaction was complete and the available soap was adsorbed on the polymer particles. This was evidenced by a rapid rise in surface tension and a breaking of the monomer emulsion.

The rate in the third stage was nearly constant suggesting a diffusion controlled process. Another possibility considered was

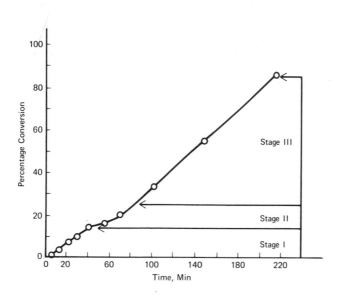

FIGURE 4.3 Conversion-time curve for the batch emulsion polymerization of VDC showing the three-stages 2.0% sodium lauryl sulfate, 0.15% catalyst (Ref. 36).

partial flocculation of the latex. The later possibility could have accounted for the observed kinetics, but the experimentally observe observed changes in particle number were too small.

Diffusion limited reaction seems more likely. A sudden loss in interfacial surface area when the monomer emulsion broke could have limited the transport of monomer through the aqueous phase. This would account for the sensitivity of the reaction to stirring rate. Evans *et al.* suggested that diffusion control was due to the low solubility of VDC in the aqueous phase but this interpretation does not seem feasible in view of the rather high solubility of VDC in H_2O reported by Wiener.

Nonetheless, deviations from normal behavior were associated with high reaction rates at higher conversion levels. No abnormalities were observed either in the seed step or in slow reactions. This raises an interesting point: How can VDC, which does not dissolve in its polymer, follow Smith-Ewart kinetics? One answer referred to earlier is that kinetic studies cannot distinguish between surface and swollen particle mechanisms when the particle size is small. But another possibility is that the polymer forms in an amorphous condition in the micelles. In support of this conclusion, Moll[39] reported that in some of his runs, the latex was still amorphous at

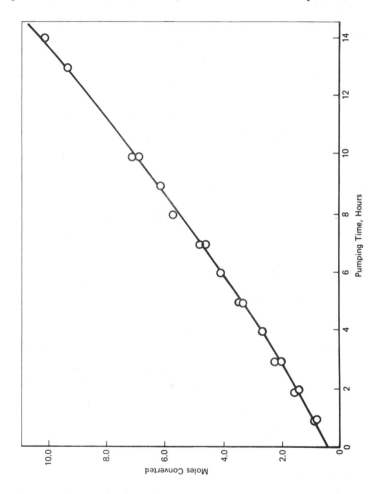

FIGURE 4.4 Emulsion polymerization of VDC with preformed seed; conversion-time curve for a flooded reaction (Ref. 28).

the end of the reaction. VDC should dissolve in amorphous PVDC because the incompatibility is due to crystallization, not unfavor-able polymer-monomer interaction.[20] Consequently, if the particles are amorphous, they should also be swollen with monomer.

The studies of Hay et al. indicate that the simple batch reaction may be too complicated to analyze quantitatively. It is possible, however, to follow seeded reaction kinetics as suggested earlier. The particle formation and growth stages have been isolated experimentally in Saran polymerization, as well as in many other cases. A "seed" is formed and then fresh monomer is added and the seed is grown to larger size. The monomer can be added in one shot or pumped into the reactor continuously at a controlled rate. If the emulsifier level is maintained between certain limits, the reaction will proceed without flocculation or formation of new particles.

An example of this type of experiment is shown in Figure 4.4. A 500 Å seed was prepared and then grown to 1500 Å by subsequent addition of monomer.[28] The monomer was added at a rate sufficient to keep excess monomer in the kettle at all times. The seed latex was checked and found to be crystallized before the growth stage. The data fit a smooth curve in a conversion time plot and show no evidence of a discontinuity. An analysis of the data using the surface model described earlier suggest that this is a surface growth reaction.

REFERENCES

1. A. D. Jenkins, In G. E. Ham (ed.), *Vinyl Polymerization*, Vol. 1, Part I, M. Dekker, N.Y., Chapter 6 (1967).
2. H. S. Mickley, A. S. Michaels and A. L. Moore, *J. Polym. Sci.*, **60**, 121 (1962).
3. L. H. Peebles, Jr., in G. E. Ham (ed.), *Copolymerization*, Interscience, N.Y., Chapter IX (1964).
4. W. M. Thomas, *Fortschr. Hochpolym.-Forsch.*, **2**, 401 (1961).
5. W. I. Bengough and R. G. W. Norrish, *Proc. R. Soc.*, **A200**, 301 (1950).
6. M. Magat, *J. Polym. Sci.*, **16**, 191 (1955).
7. W. M. Thomas and J. J. Pellon, *J. Polym. Sci.*, **8**, 329 (1954).
8. A. Schindler and J. W. Britenbach, *Ric. Sci.*, **25A**, 34 (1955).
9. V. V. Mazurek, *Vysokomol. soyed.*, **8**, 1174 (1966).

10. J. D. Cotman, Jr., M. F. Gonzalez and G. C. Claver, *J. Polym. Sci.,* A-1, **5**, 1137 (1967).
11. O. G. Lewis and R. M. King, Jr., *Adv. Chem. Ser.,* **91**, 25 (1969).
12. R. A. Wessling and I. R. Harrison, *J. Polym. Sci.,* A-1, **9**, 3471 (1971).
13. R. A. Wessling, J. H. Oswald and I. R. Harrison, *J. Polym. Sci. Phys.,* **11**, 875 (1973).
14. J. D. Burnett and H. W. Melville, *Trans. Faraday Soc.,* **46**, 976 (1950).
15. E. J. Arlman and W. M. Wagner, *Trans. Faraday Soc.,* **49**, 832 (1953).
16. C. H. Bamford and A. D. Jenkins, *Proc. R. Soc.,* **A216**, 515 (1953).
17. W. I. Bengough and R. G. W. Norrish, *Proc. R. Soc.,* **A218**, 149 (1953).
18. C. H. Bamford, A. D. Jenkins, M. C. R. Symons and M. G. Townsend, *J. Polym. Sci.,* **34**, 181 (1959).
19. C. E. Bawn, T. P. Hobin and W. J. McGarry, *J. Chim. Phys.,* **56**, 791 (1959).
20. D. E. Harmer and J. A. Raab, *J. Polym. Sci.,* **55**, 821 (1961).
21. G. Talamini and G. Vidotto, *Chim e Ind.,* **46**, 371 (1964).
22. R. A. Wessling, *J. Appl. Polym. Sci.,* **14**, 1531 (1970).
23. N. G. McCrum, B. E. Read and G. Williams, *Anelastic and Dielectric Effects in Polymeric Solids,* Wiley, N.Y., Chapter 11 (1967).
24. A. Bailey and E. H. Everett, *J. Polym. Sci.,* A-2, **7**, 87 (1969).
25. W. I. Bengough and R. G. W. Norrish, *Proc. R. Soc.,* **A218**, 155 (1953).
26. H. W. Melville and L. Valentine, *Proc. R. Soc.,* **A200**, 358 (1950).
27. J. W. Vanderhoff, in G. E. Ham (ed.), *Vinyl Polymerization,* Vol. 1, Part 2, M. Dekker, N.Y., Chapter 1 (1970).
28. R. A. Wessling and D. S. Gibbs, *J. Macromol. Sci. Chem.,* A7, 647 (1973).
29. B. M. E. Vanderhoff, *Adv. Chem. Ser.,* **34**, 1 (1962).
30. J. L. Gardon, *J. Polym. Sci.,* A-1, **6**, 623 (1968).
31. S. S. Medvedev, in *International Symposium on Macromolecular Chemistry,* Pergamon Press, N.Y., pp. 174-190 (1959).
32. L. C. Rubens and R. F. Boyer, in R. H. Boundy and R. F. Boyer (eds.), ACS Monograph No. 115, Reinhold Publishing Corp. (1952).
33. J. J. P. Staudinger, *Brit. Plastics.* **20**, 381 (1947).
34. H. Wiener, *J. Polym. Sci.,* **7**, 1 (1951).
35. G. V, Tkachenko and P. M. Khomikovskii, *Dokl. Akad. Nauk SSSR,* **72**, 543 (1950); C.A., **44**, 10373 (1950).
36. P. M. Hay et al., *J. Appl. Polym. Sci.,* **5**, 23 (1961).
37. J. C. Light, et al., *J. Appl. Polym. Sci.,* **5**, 31 (1961).
38. C. P. Evans, et al., *J. Appl. Polym. Sci.,* **5**, 39 (1961).
39. H. W. Moll, The Dow Chemical Company, unpublished results.

CHAPTER 5

Structure

PVDC has a symmetrical repeat unit which allows the polymer chains to pack efficiently in the solid state; consequently, it is highly crystalline. Comonomer units break up the regularity of the chain and interfere with the ability to crystallize. Since the properties of Saran are to a great extent controlled by crystallinity, copolymerization can be used to advantage in designing polymers for specific uses. Therefore, an understanding of copolymer composition and microstructure is even more important than an understanding of the homopolymer structure. But since the latter serves as a benchmark to which copolymer structures can be compared, the discussion of structure will start with a detailed examination of what is known about the PVDC chain structure and how the chains are arranged in the solid state.

Our knowledge in this area is not complete. The structure of the repeat unit is well established. The crystal structure has been studied extensively but is still not completely solved. Relatively little has been reported on such important aspects of structure as molecular weight distribution, branching, and solid state morphology.

5.1 ANALYSIS OF PVDC STRUCTURE

The polymer chain has a head-to-tail structure as shown below

$$-CH_2-CCl_2-CH_2-CCl_2-CH_2-CCl_2-$$

The repeat unit is symmetrical, eliminating any possibility of steroisomerism. Defects in the ideal structure described above

55

could include head-to-head units, branch points or unsaturated units caused by dehydrochlorination.

Other degradation reactions that do not cause chain scission could also introduce defects. Possibilities include a variety of ill-defined oxidation and hydrolysis reactions which introduce oxygen into the polymer.

This is evident from the infrared spectra of PVDC which often shows bands due to unsaturation and carbonyl absorption. The slightly yellow tinge of many of these polymers comes from the same source; a very pure sample of the polymer is colorless.

Elemental analyses for chlorine content in the polymer are close to theoretical (73.14%) suggesting that dehydrochlorination during polymerization can only take place to a small extent.

The first direct evidence for head-to-tail addition in PVDC came from x-ray diffraction studies. The early investigators Staudinger,[1] Frevel[2] and Fuller[3] observed that the diffraction patterns were characteristic of a highly crystalline polymer implying that PVDC had a regular structure. Further analysis of the diffraction patterns led to a correct assignment of the repeat unit.

The head to tail arrangement was later confirmed by an analysis of the IR spectra of PVDC. An example of a typical spectrum obtained from a highly crystalline powder is shown in Figure 5.1. The spectra from solid specimens are sensitive to both crystallinity and orientation. Bands at 753 and 883 cm^{-1} are absent in the amorphous polymer. The strong doublet at 1046 and 1071 cm^{-1} is a characteristic feature of the homopolymer and crystalline copolymers. The carbon-chlorine stretching region is another identifying feature (500-700 cm^{-1}).

In order to interpret the structure of PVDC correctly, spectroscopists have been concerned with identifying the major bands. Although earlier workers made contributions, the first reasonably successful attempt at a complete assignment was reported by Krimm and Liang[4] in 1956. This was followed by the more detailed investigations of Narita et al.[5,6] In order to elucidate the structure of PVDC more reliably, they made comparisons with deuterated PVDC, polyvinylidene bromide and various vinylidene chloride-vinyl chloride copolymers. Both groups utilized polarized IR on oriented specimens as well.

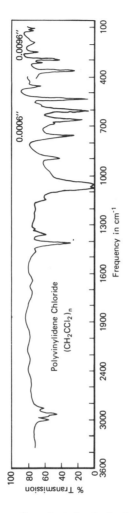

FIGURE 5.1 Infrared spectrum of unoriented polyvinylidene chloride (Ref. 4).

In 1960, Krimm[7] reviewed the available data and made some adjustments in the assignments. A further modification was proposed recently based on Raman spectra.[8,9] Both IR and Raman spectroscopy support the assignment of a head-to-tail structure to PVDC.

Further confirmation of the head-to-tail structure came out of studies of the NMR spectra of PVDC.[10-13] A typical spectrum shows only a single peak at τ = 6.18, assigned to the methylene group proton. All protons in a head-to-tail structure occupy equivalent positions and should appear as a single resonance line, in agreement with the observed spectrum.

Another mode of addition or extensive branching would produce non-equivalent hydrogens and a more complicated spectrum. If such structures are present their concentrations are below the limit of resolution of this method ($<$ 1%). The head-to-tail structure has also been supported by degradation and pyrolysis studies.[14,15]

5.2   MOLECULAR WEIGHT

While the chain structure is well established from the point of view of chemical bonding, relatively little is known about chain structure in the larger sense (chain length, branching, etc.). Absolute molecular weight measurements by light scattering have been only recently reported.[16] The results were used to establish intrinsic viscosity-molecular weight relationships. But problems relating to polydispersity, chain branching, etc., have not yet been addressed.

In view of its polymerization chemistry, polymers of VDC should show the same range of molecular weights and the same type of distributions observed with other sterically hindered monomers such as methyl methacrylate. Studies on copolymers indicate that this is the case.[17] The ratio $\bar{M}_w / \bar{M}_n$ falls in the range of 1.5 to 2.0.

A limited amount of information on chain length and branching has been obtained from viscosity measurements. In some early studies, McIntire[18] compared viscosity/molecular weight relationships for PVDC and PVC and concluded that the former was essentially linear whereas the latter was highly branched. The high crystallinity of PVDC also suggests a relatively linear chain.

Polymers with intrinsic viscosities ranging from 0.01 to $>$ 2 dl/g can be readily prepared by free radical polymerization. These data indicate that degrees of polymerization from 100 to over 10,000 can be realized. This is more or less typical for free radical polymerization.

5.3 CRYSTAL STRUCTURE

The arrangement of the polymer chains in the crystalline state has been studied extensively by x-ray diffraction techniques. Several unit cells have been proposed,[2,19,20] but the exact arrangement of the chains within the cell is not known with certainty. Unit cell data are collected in Table 5.1. Agreement on unit cell density and repeat distance in the chain direction is good.

TABLE 5.1

Crystallographic Data for Polyvinylidene Chloride

Crystal System	Space group	A	B	C	$\beta_{(deg)}$	Monomer per cell	Density g/c^3
Monoclinic		13.69Å	4.67Å	6.296 Å	123.8	4	1.949
Monoclinic		22.54	4.68	12.53	84.2	16	1.959
Monoclinic	C2-2	6.73	4.68	12.54	123.6	4	1.96

The calculated densities are somewhat higher than the range observed experimentally, 1.80-1.94 g/c^3. But this is normal for semi-crystalline polymers because 100% crystallinity cannot be achieved.

Unit cell data have been obtained primarily from x-ray studies on oriented fibers. A typical example of a pattern from a highly oriented fiber is shown in Figure 5.2. Okuda[20] has indexed some of the important reflections in both PVDC and copolymers with VC. The spacings listed in Table 5.2 indicate the structure of PVDC is retained in the copolymers.

Fiber patterns provide only a limited amount of data for structure analysis. In addition, the quality of the data is poor due to the strong absorption of the x-rays (CuKα) by PVDC. In spite of these difficulties, several workers have tried to carry out a structure analysis.

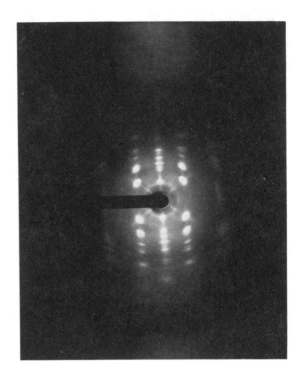

FIGURE 5.2 X-ray diffraction pattern from a highly oriented PVDC fiber.

(Courtesy D. R. Carter, The Dow Chemical Company)

5.3.1 Conformations

One feature of the structure is readily decided from the fiber pattern spacings. The repeat distance along the chain, 4.68 Å involves two monomer units. This is significantly less than the distance corresponding to a planar zig-zag conformation. Therefore, the polymer must be in some other disposition. A number of conformations have been proposed, but there is no agreement as yet on which is correct. Frevel[2] proposed a 2_1 helix with expanded bond angles. This structure gave the best agreement with his x-ray diffraction

TABLE 5.2

Relationship between Compositions and Observed Spacings of
VDC–VC Copolymers (Ref. 20)

Composition (molar fraction of VDC)	100 + $\overline{1}$02	002	$\overline{1}$04	200 + $\overline{2}$04	$\overline{1}$05 + $\overline{2}$05	005 + $\overline{3}$05	020
				Observed spacings, Å			
1.000 (PVDC)	5.61	5.35	3.11	2.804	2.440	2.081	2.343
0.905	5.61	5.28	3.10	2.804	2.440	2.089	2.343
0.820	5.63	5.28	3.11	2.809	2.443	2.086	2.343
0.745	5.61	5.28	3.11	2.809	2.443	2.086	2.343
0.685	5.63	5.28	3.11	2.809	2.450	2.089	2.350
0.603	5.63		3.13	2.809	2.450	2.089	2.343
0.560	5.57			2.809			2.350

data. The conformation is sketched out in Figure 5.3. Fuller[3] suggested a staggered planar zigzag, but noted that a helical structure would also be consistent with his x-ray data. DeSantis et al.[21,22] conclude, from x-ray data, a conformational energy analysis, and optical Fourier transforms of the various models, that the helical conformation is correct. Okuda[23] has also carried out a partial structure analysis using x-ray diffraction methods and concludes that the TGTG' (a modified version of the Fuller structure) conformation proposed by Miyozawa and Ideguchi[24] is more likely. The latter authors deduced this conformation from a study of the infrared spectra. Other infrared spectroscopists have reached a similar conclusion.[7] Additional support for this conformation has come from a recent analysis of the laser-excited Raman Spectra of PVDC.[8]

The spectroscopic determinations of chain conformation are based on the assignment of the four strong bands believed to be C–Cl stretching modes. Hendra et al.[8] identify these as shown in Table 5.3. Three of these bands disappear in the molten polymer indicating that they are associated with the crystalline phase. The band at 530 cm^{-1}, however, is observed in the amorphous polymer.

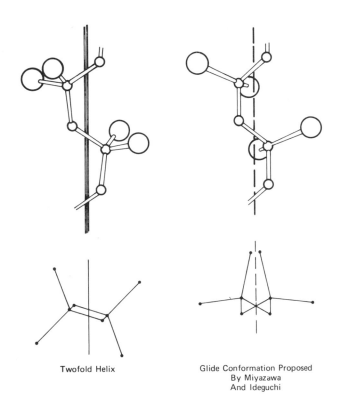

Twofold Helix

Glide Conformation Proposed
By Miyazawa
And Ideguchi

FIGURE 5.3 Chain model of poly (vinylidene chloride) (Ref. 22).

Parallel alignment of oriented specimens with a polarized beam
leads to strong absorption at all four wavelengths; perpendicular
orientation greatly reduces the absorbance. This indicates that the
C–Cl bonds are perpendicular to the chain axis.

The correct assignment of the C–Cl stretching modes is critical
to the determination of conformation. The *TGTG'* structure should
show four bands and a helical structure only three. Most spectroscopists have agreed on the presence of four such bands but
additional work is needed to be completely certain.

5.3.2 Comparison of vinylidene-type structures

Further information about the chain conformation of PVDC can be gleaned from a comparison of this polymer with other 1,1 di-substituted ethylene polymers. These include polyvinylidene bromide (PVDB), polyisobutylene (PIB) and polyvinylidene fluoride (PVDF). Selected data are shown in Table 5.4. PVDB is isomorphous with PVDC.[18] PVDF (modification II) has a structure also closely related to that of PVDC.[25,26] But PIB does not conform to the pattern. Its identity period is much longer.[27]

TABLE 5.4
Comparison of Structures of Vinylidene Polymers

Repeat unit	Repeat distance, Å	Chain conformation
$-CH_2CF_2-$	4.62	$TGTG'$
$-CH_2-CCl_2-$	4.68	$TGTG'$ (?)
$-CH_2-CBr_2-$	4.77	—
$-CH_2-C(CH_3)_2-$	18.6	8/3 helix

The most recent studies on the conformation of PIB support an 8/3 helix with an expanded $C-CH_2-C$ bond angle of $123°$.[28] This suggests that one of the principal objections to the Frevel structure, the large bond angle, may not be warranted. Expanded angles that reduce steric repulsions between substituents in a steric-ally crowded chain appear to be energetically favored. In comparing the series of vinylidene polymer structures, it is apparent that the repeat distance increases gradually as the size of the substituent increases from F to Br; methyl substitution, however, causes a change in conformation probably because the steric repulsions become intolerable in the $TGTG'$ conformation.

The $TGTG'$ conformation, favored by spectroscopic data, has several objectionable features. The structure does not minimize the conformational energy but instead brings two of the chlorine sub-stituents into close proximity. This would be acceptable if the $TGTG'$ chain were able to pack efficiently, but the packing is actually quite poor.[26] Until further data are available the tentative assignment of the $TGTG'$ conformation for PVDC seems reasonable. But it should be noted that the Frevel structure has been used successfully to interpret both degradation[29] and fracture[30] of PVDC lamellar crystals.

As was noted already for the VC case, the unit cell dimensions of PVDC do not appear to be affected by the presence of co-monomer units in the chain. However, studies of melting point depression suggest that VC may cocrystallize to a limited extent and could have an effect. Other comonomers such as the acrylates are effectively excluded from the crystal and should not influence the crystal structure.

5.4   COPOLYMER MICROSTRUCTURE

The effect of comonomer on the solid state properties of PVDC is determined by the way it is placed in the polymer, i.e., on the microstructure.

There are three aspects of copolymer structure to consider: the average composition of a polymer chain; the sequence distribution in a given chain; and finally the composition and sequence distribution in a high conversion copolymer. These quantities can be

TABLE 5.5

List of Saran Copolymers Systems for which the Microstructure has been Analyzed

Comonomer	Method	Reference
VC	NMR	10-13, 35
VC	IR	5, 36, 37
VC	Raman	9
VC	Pyrolysis	15
VC	Density gradient	38
Isobutylene	NMR	39-42
Isobutylene	IR	43
Vinyl acetate	NMR	33, 44
Methyl methacrylate and other methacrylates	NMR	45, 46
Methacrylonitrile	NMR	47

derived from the copolymerization theory.[31-34]

Microstructure can be determined by a number of techniques including NMR, IR, fractionation and pyrolysis. The NMR method is most widely used. The systems studied and the methods are listed in Table 5.5.

REFERENCES

1. H. Staudinger and W. Feisst, *Helv. Chim. Acta,* **13**, 805 (1930).
2. L. K. Frevel, The Dow Chemical Company, unpublished results.
3. C. S. Fuller, *Chem. Rev.,* **26**, 143 (1940).
4. S. Krimm and C. Y. Liang, *J. Polym. Sci.,* **22**, 95 (1956).
5. S. Narita, S. J. Chinohe and S. Enomoto, *J. Polym. Sci.,* **37**, 251 (1959).
6. *Ibid.,* **37**, 263 (1959).
7. S. Krimm, *Fortschr. Hochpolym. Forsch,* **2**, 51 (1960).
8. P. J. Hendra and J. R. Mackenzie, *Spectrochim. Acta,* **25A**, 1349 (1969).
9. M. R. Meeks and J. L. Koenig, *J. Polym. Sci. A-2,* **9**, 717 (1971).
10. R. Chujo, S. Satoh and E. Nagai, *J. Polym. Sci. A,* **2**, 895 (1964).
11. J. L. McClanahan and S. A. Previtera, *J. Polym. Sci. A,* **3**, 3919 (1965).
12. U. Johnsen, *Kolloid-Z. Z. Polym.,* **210**, 1 (1966).
13. U. Johnsen, *Ber. Bunsenges. Phys. Chem.,* **70**, 320 (1966).
14. R. F. Boyer, *J. Phys. Coll. Chem.,* **51**, 80 (1947).
15. S. Tsuge, T. Okumoto and T. Takeuchi, *Makromol. Chem.,* **123**, 123 (1969).
16. K. Matsuo and W. H. Stockmayer, *Macromolecules,* **8**, 660 (1975).
17. M. L. Wallach, ACS, *Polymer Div. Preprints,* **10**, 1248 (1969).
18. O. R. McIntire, The Dow Chemical Company, unpublished results.
19. S. Narita and K. Okuda, *J. Polym. Sci.,* **38**, 270 (1959).
20. K. Okuda, *J. Polym. Sci. A,* **2**, 1749 (1964).
21. P. DeSantis, E. Giglio, A. M. Liquori and A. Ripamonti, *J. Polym. Sci. A,* **1**, 1383 (1963).
22. *Ibid., Polym. Lett.,* **4**, 821 (1966).
23. K. Okuda, private communication.
24. T. Miyazawa and Y. Ideguchi, *Polym. Lett.,* **3**, 541 (1965).
25. F. J. Boerio and J. L. Koenig, *J. Polym. Sci.,* **7**, 1489 (1969).
26. W. W. Doll, The Polymorphism and Isomorphism of Poly(vinylidene fluoride), Ph.D. Thesis, Case-Western Reserve University (1970).
27. C. W. Bunn and D. R. Homes, *Disc. Faraday Soc.,* **25**, 95 (1958).
28. G. Allegra, E. Beneditti and C. Pedone, *Macromolecules,* **3**, 727 (1970).
29. I. R. Harrison and E. Baer, *J. Coll. Interface Sci.,* **31**, 176 (1969).
30. A. Bailey and D. H. Everett, *J. Polym. Sci.,* **A-2, 7**, 87 (1969).

31.   T. Alfrey, Jr., J. J. Bohrer and H. Mark, *Copolymerization*, Interscience, N.Y. (1952).
32.   G. E. Ham (ed.), *Copolymerization*, Interscience, N.Y. (1964).
33.   Y. Yamashita, K. Ito, H. Ishii, S. Hoshino and M. Kai, *Macromolecules*, 1, 529 (1968).
34.   V. E. Meyer and G. G. Lowry, *J. Polym. Sci. A*, 3, 2483 (1965).
35.   S. Enomoto and S. Satoh, *Kolloid-Z. Z. Polym.*, 219, 12 (1967).
36.   S. Enomoto, *J. Polym. Sci.*, 55, 95 (1961).
37.   H. Germar, *Markomol. Chem.*, 84, 36 (1965).
38.   R. F. Boyer, R. S. Spencer and R. M. Wiley, *J. Polym. Sci.*, 1, 249 (1946).
39.   T. Fisher, J. B. Kinsinger and C. W. Wilson, *Polym. Lett.*, 4, 379 (1966).
40.   *Ibid.*, 5, 285 (1967).
41.   U. Johnsen and K. Kolbe, *Kolloid-Z. Z. Polym.*, 232, 712 (1969).
42.   K. H. Hellwege, U. Johnsen and K. Kolbe, *Kolloid-Z. Z. Polym.*, 214, 45 (1966).
43.   U. Johnsen and W. Lesch, *Kolloid-Z. Z. Polym.*, 232, 863 (1969).
44.   U. Johnsen and K. Kolbe, *Kolloid-Z. Z. Polym.*, 216-217, 97 (1967).
45.   K. Ito, S. Iwase and Y. Yamashito, *Makromol. Chem.*, 110, 233 (1967).
46.   Y. Yamashito, K. Ito, S. Ikuma and H. Kada, *Polym. Lett.*, 6, 219 (1968); 6, 227 (1968).
47.   R. E. Block and H. G. Spencer, *J. Polym. Sci. A-2*, 9, 2247 (1971).

CHAPTER 6

The Crystalline State

6.3 MEASUREMENT OF CRYSTALLINITY

6.1.1 General discussion of crystallinity

The stable state of PVDC at ambient temperatures is the crystalline state. In most cases, the polymer crystallizes as it is formed in the polymerization reaction and can be isolated as a highly crystalline powder. Copolymers may also form as crystalline powders depending on the type and concentration of comonomer. The tendency to crystallize during polymerization falls with increasing comonomer content until a point is reached where the polymer is completely amorphous. At some intermediate level, the copolymer may crystallize but with difficulty. In such cases, it remains amorphous during preparation but may crystallize during isolation or as a result of annealing or mechanical working.

Permanently amorphous Saran copolymers, like other non-crystalline polymers, have a relatively unordered structure in the solid state. Both PVDC and crystalline copolymers can be melted and quenched to a similar amorphous state. But they soon re-crystallize unless kept at temperatures well below 0 °C. The recrystallized polymer bears little resemblence to the "as polymer-ized" powder. It is substantially lower in crystallinity and has a spherulitic morphology. If the quenched polymer is cold drawn it crystallizes into an oriented fibrillar structure.

The crystalline state is normally characterized by a single number, the degree of crystallinity (or % crystallinity).

Values obtained for % crystallinity depend on the measuring technique. In the commonly used density method, it is given by

Equation (6.1):

$$\% \text{ Cryst. } = \frac{V - Va}{Vc - Va} \times 100 \, , \tag{6.1}$$

where

V = specific volume of sample

Va = specific volume of amorphous polymer

Vc = specific volume of 100% crystalline sample (usually based on unit cell volume).

This measure is based on the assumption of a two phase model for semi-crystalline polymers. This model is a gross over-simplification of the actual morphology of crystalline polymers. Therefore, percent crystallinity is, at best, an estimate of the average degree of order in a polymer. It tells nothing of the detailed structure.

The importance of morphology in determining the properties of a Saran copolymer is clearly illustrated by some early experiments by Lowry and Wiley.[1] They compared samples that were melted and quenched to 10 °C and subjected to various annealing schedules before crystallization at 140 °C. Samples aged several days at 10 °C and then heated to 140 °C crystallized very quickly but remained transparent and ductile. Samples that were quenched and quickly raised to a high temperature became turbid and brittle. Yet both materials had the same density, and according to Equation (6.1) should have the same % crystallinity.

The difference in behavior of these materials is related to differences in morphology. The clear sample probably contained many very small spherulites; the cloudy speciment, few but relatively large spherulites.

6.1.2   Measurement of crystal dimensions

Ideally, we would like to know not only spherulite size but also the structure of the spherulite, i.e., to have a description of the basic crystalline units or crystallites. For example in a polymer with a lamellar morphology we would like to know the lamellar thickness as well as the percent crystallinity and spherulite size.

Various techniques have been applied to elucidate the crystallite size in melt crystallized Saran. Wiley and coworkers used a polarizing microscope to observe changes in texture. At the same time,

Frevel applied the x-ray line broadening method.[2] His analysis indicated that the basic unit in melt crystallized Saran was a platelet with thickness of 20-30 Å in the chain direction and 200-500 Å in width. Due to the relatively low crystallinity of these samples, the Dow workers recognized that the crystalline regions must be separated by amorphous or disordered regions.

They favored the picture of a continuous interpenetrating network of crystalline and amorphous regions as being best able to explain the properties of Saran.[1-3] This model is remarkably close to modern views of the morphology of crystalline polymers.[4-6]

More recent studies by Okuda,[7] Kockott[8] and Bailey and Everett[9] have confirmed the findings of the early workers. Working with both oriented specimens and lamellar crystals as well as the virgin polymers, Okuda found crystalline regions in VDC/VC copolymers to be of the order of 40 Å in the chain direction. His studies also indicated that a substantial fraction of the polymer was in an unordered state in regions between crystals. This would require that amorphous regions of comparable dimensions must separate the crystalline regions.

Kockott[8] studied the dimensions of crystallites in the same copolymer system but with samples of narrow distribution of chemical composition. He analyzed the dependence of the degree of crystallinity on copolymer composition in terms of Flory's theory of copolymer crystallization.[10] He compared the derived minimum sequence length to values calculated from low angle x-ray scattering. The experimental values he obtained were comparable to Frevel's estimate and agreed poorly with the theoretical estimates. All three studies are in agreement that the dimensions of the crystals in the chain direction are very small. The agreement, in fact, is much better than should be expected because of the great differences in the method of preparation of the various specimens. Frevel examined melt crystallized specimens; Kockott used samples crystallized from solution; and Okuda studied samples that were extruded, quenched and cold drawn.

6.1.3 Degree of crystallinity

There are also significant differences in the range of crystallinities reported in the above mentioned studies. Frevel measured values of approximately 75% for "as polymerized" PVDC and high VDC copolymers. After melting and recrystallization, he found values ranging from 20-50% depending on temperature of crystallization. Okuda observed values of 19-43% on the "as polymerized" powders with crystallinity increasing with VDC content. Over the same composition range, Kockott calculated values of 5-55% for recrystallized specimens.

These authors used x-ray techniques. Therefore, the values obtained are influenced by the degree of order in the amorphous fraction and the degree of perfection in the crystalline regions as well as the method used to analyze the scattering factor in terms of contributions from the amorphous and the crystalline fraction of the sample.[11,12]

Problems in measuring percent crystallinity are common to all polymers but are particularly aggravating in the case of Saran. Difficulties arise not only when possible order in the amorphous state is taken into account, but also because conversion to percent crystallinity requires the definition of a completely amorphous or completely crystalline state to which experimental data can be compared. In the case of PVDC, the problem is illustrated by the use of calorimetric methods that depend on a knowledge of $\Delta H\mu$, the heat of fusion per mole of perfect crystalline units and volumetric measurements which depend on a knowledge of the density of the pure amorphous polymer.

Early measurements on melt crystallized Saran copolymers indicated that $\Delta H\mu$ was less than 0.5 kcal/mole.[3] Okuda[7] reported a value of ~ 0.63 kcal/mole for suspension polymerized PVDC. Using the x-ray value of 43% crystallinity, he obtained a value of 1.5 kcal/mole for $\Delta H\mu$, in agreement with the results of T_M depression studies of random copolymers and in solution. However, Kockott[8] reported a value of $\Delta H\mu$ = 1.9 kcal/mole from T_M depression in VDC/VC polymers.

Directly measured heats of fusion on highly crystalline PVDC fall in the range of 1.0 to 1.1 kcal/mole.[13,14] This establishes a lower limit for $\Delta H\mu$. Experimental values obtained recently by melting point depression studies in various solvents[15] support a value of $\Delta H\mu$ around 1.5 kcal/mole. Assuming that the fraction of crystalline material is proportional to the ratio of the measured heat of fusion to $\Delta H\mu$, levels of crystallinity of the order of 70% in "as polymerized" PVDC are obtained.

If order exists in the amorphous regions, this approach would give too high a value for the degree of crystallinity. The available data, however, suggest that any error from this source is small. Wiley[16] measured densities of 1.757 g/c^3 at 25 °C on quenched and apparently amorphous samples of a high VDC content co-polymer ($>$ 95% VDC). Eykamp et al.[17] reported a value 1.7754 g/c^3 for the density of amorphous PVDC, obtained by extrapolation of amorphous copolymer densities.

In addition, crystallinities calculated from x-ray (59%), density (70%), and calorimetric data (66%), on the same batch of low conversion mass PVDC[20] are in reasonable agreement.

Levels of crystallinity as high as 85% (by density) have been observed on highly crystalline PVDC samples.[15] Even taking into account the difference between methods, this suggests that the levels of crystallinity reported by Kockott[8] as the maximum attainable were on the low side. It follows, therefore, that the melting points he obtained were not equilibrium values.

Maximum crystallinities are usually obtained by slow crystallization from the melt near the melting point. In such experiments, the observed melting point increases with crystallization temperature. But even under these conditions, it is virtually impossible to establish equilibrium because of the size and chain like nature of polymer molecules. These experiments cannot be carried out with PVDC because of severe degradation above 130 °C.[18] To avoid this problem, Kockott[8] crystallized his samples from chlorobenzene at 120 °C. But even under these conditions, equilibrium is not established and the stability of the crystalline phase will be less than that of a hypothetical perfect crystalline sample.

6.2 MORPHOLOGY

6.2.1 Morphology of "as polymerized" PVDC

However it is measured, the degree of order in a particular sample depends not only on composition but also on method of preparation and subsequent thermal history. The % crystallinity of "as polymerized" PVDC is dependent on polymerization conditions. This appears to be related to polymer morphology. The highest values are found for samples prepared in a low conversion mass or slurry reaction.

The morphology of these materials is complex. Bailey and Everett[9] were able to isolate ribbon like lamellae and platelets from PVDC powders. These crystals were 90-100 Å thick. They assumed that the crystals had a regularly folded lamellar structure as is commonly observed in crystalline polymers. With this assumption, they were able to correlate observed faces and fracture surfaces with low index planes in the lattice.

Bort and Arzhakov[19] observed two types of particles: globules formed by surface polymerization and platelets formed by precipitation of the polymer from solution. The latter showed a well-defined screw dislocation growth habit.

The powders obtained by mass polymerization are very porous as suggested by low bulk densities. Surface areas of the order of 90 m^2/g (N_2 adsorption) have been measured on low conversion samples.[20] Particle size increases and porosity decreases with increasing conversion.

The complexity of the structure of these particles is illustrated by the micrographs in Figure 6.1. Mass polymerization to low conversion is very similar to slurry polymerization in a solvent medium that does not dissolve PVDC. In the latter case, however, the morphology of the "as polymerized" polymer is strongly influenced by the interaction between polymer and solvent.[20]

As also shown in Figure 6.1, particles formed in a non-swelling medium such as hexane or methanol resemble those isolated from mass reactions at low conversion. A completely different morphology develops when polymerization is carried out in a medium that swells the polymer but does not dissolve it. Typical systems include dilute solutions of VDC (10-20%) in dioxane or cyclohexanone. Examples of this type of morphology are also shown in Figure 6.1.

The mechanism by which these particles form must involve rapid crystal growth under conditions of very high supercooling. The favored habit in this situation is the curved lamella.[21] This is substantiated by the fact that the particles of very similar morphology can be grown by dissolving PVDC in the same solvents and

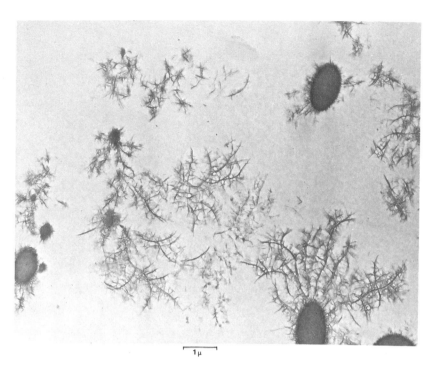

1μ

FIGURE 6.1 Morphology of "as polymerized" PVDC; transmission electron
micrographs of sectioned particles.

(a) Polymerized in mass.

recrystallizing it rapidly at the polymerization temperature used for
the original experiments. However, kinetic data indicate that VDC
polymerizes in dioxane by the normal heterogeneous reaction
mechanism, so it is possible that some of the crystal growth involves
polymerization on the existing crystal surface.

There is a smooth variation in morphology as the activity of
the solvent is varied. A solvent like benzene which is intermediate
in solvent power yields doubly curved but less highly branched
lamellae. Particles of identical appearance to those formed in benzene

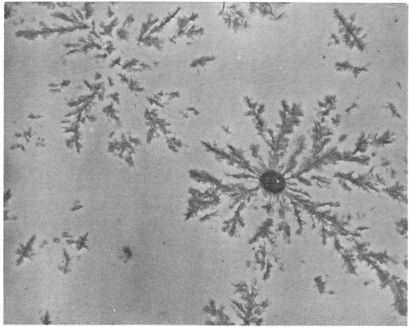

FIGURE 6.1

(b) In cyclohexane.

can be generated by using the appropriate mixture of cyclohexane
and dioxane to get the same solvent power.

6.2.2 Crystals from solution

When PVDC is crystallized slowly from very dilute solution, it forms
folded chain lamellar crystals. A particularly good example is shown
in Figure 6.2. Okuda[7] has confirmed the chain orientation by
electron diffraction studies but found the crystals he prepared to be
highly twinned. Recent studies have confirmed that this is a characteristic feature of PVDC crystals.[23]

Crystals with lamellar thicknesses ranging from 40-80 Å can be

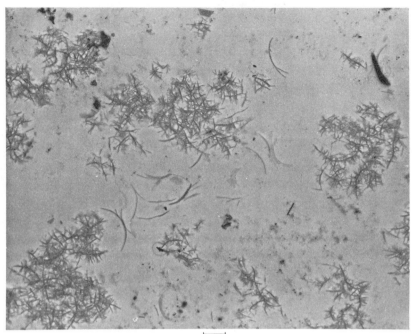

FIGURE 6.1

(c) In benzene

grown depending on solvent and temperature. The quality varies considerably. In polar solvents, like cyclopentanone, the edges are ragged and poorly defined.[14,22] Better crystals with flat faces and sharp angles can be obtained from non polar solvent like 1,2-dibromoethane.[23] Generally, however, single well defined platelets like those that have been reported for polyethylene cannot be obtained. Instead, highly branched three dimensional clusters are developed.

The branching angle shows significant regularity, usually falling in the range of 65-70°. This angle is related to the location of one

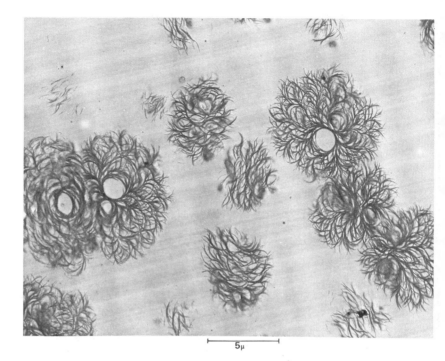

5μ

FIGURE 6.1

(d) In propylene oxide.

of the twin planes in the crystal.[23]

As is characteristic of other polymers, the morphology of PVDC crystallized from solution changes from lamellar at low concentrations (< 1%) to spherulitic at high concentrations (> 20%).[24] The ability to crystallize in either habit is impaired by copolymerization. As the comonomer content is raised, lamellae become less distinct in form; eventually a level is reached where the polymer precipitates as an irregular cluster. Solutions of copolymers with even lower crystallizability tend to gel without forming a distinct precipitate at all.

PVDC Lamellar Crystals

$1\,\mu$

FIGURE 6.2 Lamellar crystals of PVDC grown in 1,2,4-trichlorobenzene at 125 °C.

6.2.3   Spherulitic morphology

When PVDC is crystallized from the melt or concentrated solution under quenching conditions, the crystalline regions are too small to resolve under an optical microscope. It is not certain if this morphology is spherulitic or has a liquid-crystalline character like quenched polypropylene.[5] Large spherulites develop at temperatures above 100 °C but degradation becomes a problem and it is more convenient to study plasticized polymer or high VDC copolymers at lower temperatures. An example of spherulitic plasticized PVDC is shown in Figure 6.3.

Our knowledge of the spherulitic morphology in PVDC is limited. Schurr[25] reported that PVDC spherulites had the normal type of maltese cross appearance if viewed between crossed polaroids. He reported that they were positively birefringent and from this deduced that the polymer chains were oriented in a tangential direction.

Recent measurements by Gurnee,[26] however, indicate that both positive and negative spherulites exist in PVDC depending on crystallization conditions. The same has been observed for other polymers also. This suggests that chain orientation cannot be deduced from the sign of birefringence alone.

Kockott[8] has published a micrograph of very well defined PVDC spherulites with diameters of $\sim 10\mu$. He gave few details on the preparative technique and did not study the morphology. Okuda et al.[27] have recently reported growth rates of spherulites in a VDC/VC copolymer but again gave no details about morphology.

In the absence of more information, we can only assume that PVDC spherulites do have a normal structure. It is generally believed[28] that spherulites are formed from fibrous crystals growing radially with a branching habit from a common nucleus. The fibers appear to be ribbon-like lamellar crystals in some cases. The polymer chains are oriented perpendicular to the fiber direction and, therefore arranged in a tangential fashion around the spherulite nucleus. The interlamellar regions contain amorphous polymer, chain folds, tie chains and impurities.

The spherulitic structure of PVDC is common to high VDC content copolymers. Copolymers of low crystallinity do not develop a well defined crystalline morphology. When crystallinity falls below

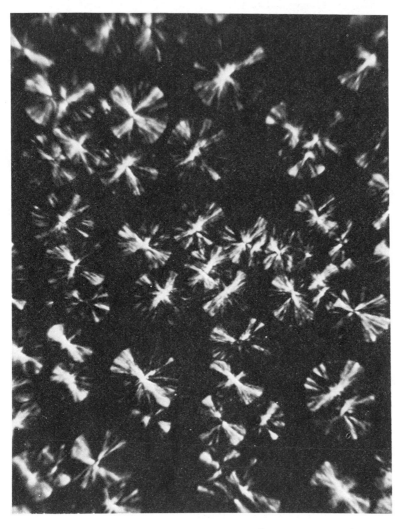

0.1 mm

FIGURE 6.3 Spherulites of PVDC grown from the plasticized melt viewed
under crossed polaroids.

~ 10%, it becomes difficult to detect though the effect on mechanical properties is pronounced. These materials behave somewhat like crosslinked rubbers. The regions of microcrystalline order are probably best described by the fringed micelle model; i.e., ordered regions in a continuous amorphous matrix, where individual chains pass through many "crystallites".

6.2.4   Fibrous and oriented structures

Polymers with a very limited crystallizability often require stetching to induce crystallization. Mechanical working greatly increases the rate of crystallization even in PVDC. Lowry and Wiley[1] found that stretching and relaxing quenched samples before crystallization markedly increased nuclei concentration in the melt. When subsequently heated, these samples crystallized more rapidly.

The higher stretch ratio (~ 150%) not only accelerated the rate of crystallization but also affected the morphology. When the relaxed fiber crystallized, it would spontaneously elongate 10-15%. The crystallized specimen showed orientation when examined in polarized light. If quenched fibers were stretched more than 150%, they crystallized during the operation. Okuda et al.[27] have studied the kinetics of this process and suggest that strain induced crystallization is preceded by orientation of the polymer chains in the melt. An analysis of the data indicate that a change in mechanism takes place between small and large stretch ratios.

The early work at Dow also showed that filaments crystallized under quenching conditions would cold draw to highly oriented fibers. The density of the cold drawn fibers was lower than that of the starting material. X-ray diffraction patterns indicated a very high degree of orientation.

In 1964, Okuda[7] reported further studies of cold drawn fibers using modern instrumental techniques such as a small angle x-ray scattering. He measured degree of crystallinity, crystallite size and long periods on a series of cold drawn and annealed fibers of varying VC content. The fibers had a high degree of chain orientation parallel to the fiber axis, ~ 94%. They exhibited the usual fiber morphology, small oriented crystalline regions regularly spaced and separated by disordered regions. Crystal dimensions and long periods are listed in Table 6.1. Annealing increased the latter but had little effect on the

TABLE 6.1

Relationship between Composition and Crystalline Parameters of
VDC–VC Copolymers (Ref. 7)

Composition (molar fraction of VDC)	Melting point °C	Crystallinity %	Drawing conditions	Average crystallite size along fiber axis Å	Long period Å
1.000 (PVDC)	195	43	Cold-drawn	45	76
			Annealed at 100 °C after being cold-drawn[a]	48	90
0.905	192	34	Cold-drawn	45	80
			Annealed at 100 °C after being cold-drawn	48	96
0.820	188	33	Cold-drawn	45	84
			Annealed at 100 °C after being cold-drawn	47	96
0.745	185	28	Cold-drawn	45	87
			Annealed at 100 °C after being cold-drawn	47	100
0.685	183	22	Cold-drawn	41	103
			Annealed at 90 °C after being cold-drawn	43	119
0.603	183	19	Cold-drawn	41	135
			Annealed at 70 °C after being cold-drawn	43	142
0.560	183	20	Cold-drawn	42	–

[a] Annealing was carried out under constant length, for 10 min, in a water bath.

crystalline regions.

Okuda and coworkers in their later study of a commercial copolymer[27] concluded that microstructure of the cold drawn fibers was mainly of the fringed micelle type with little of the folded chain lamellar character observed in single crystals.

6.3　KINETICS OF CRYSTALLIZATION

6.3.1　Crystallization of VDC–VC copolymers

The above discussion brings out the point that the morphology of melt crystallized Saran is influenced by mechanical and thermal treatment as well as copolymer composition. These same factors influence the kinetics of crystallization also. PVDC itself crystallizes readily from ∼ 0 to 175 °C. The kinetics have not been studied in detail but the maximum rate appears to fall in the range of 100-150 °C.

Copolymers crystallize more slowly over a narrower temperature range. Several systems have been studied. Most of the early work at the Dow Chemical Company was carried out with broad composition distribution copolymers. An increase in VC content was found to increase the time lag before crystallization started (induction time); it reduced both the rate of crystallization and the total level of crystallinity achieved.

Crystallization induction times for a given copolymer were found to be very temperature sensitive. The induction time was determined either optically (by the appearance of turbidity)[1] or by the density gradient technique.[29] Figure 6.4 shows the dependence of induction time on temperature.

Small amounts of plasticizer reduced the induction time at lower temperatures and shifted the minimum to a lower temperature also. The plasticizer apparently increases the mobility of polymer chains at the lower temperatures by depressing T_G.[30]

The above results have been confirmed by the recent studies of Okuda et al.[27]

6.3.2　Crystallization of acrylate copolymers

Another consequence of the effect of copolymer composition on transitions is to narrow the temperature range in which crystallization can take place. As the comonomer level is increased, the melting point of the polymer falls while the glass transition temperature, T_G, increases. The copolymer can only crystallize between these limits. Therefore, a copolymer with a high T_M and a low T_G should crystallize most easily.

This effect was clearly illustrated in the kinetic studies on

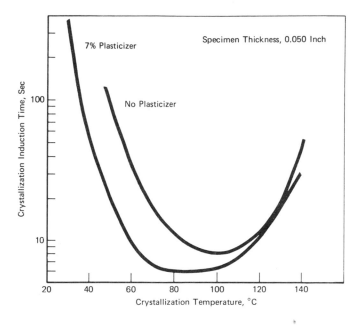

FIGURE 6.4 Typical crystallization induction period curves for a vinylidenc chloride-vinyl chloride copolymer containing about 10% VC; Broad distribution (Ref. 3).
A — No plasticizer; B — 7% plasticizer; specimen thickness, 0.050 inch.

acrylate copolymers by Riser and Witnauer.[31] On three copolymers each containing 90 mole % VDC and 10 mole % of an n-alkyl acrylate. The length of the alkyl group was varied from butyl to octyl to octadecyl. Crystallization isotherms were measured by monitoring changes in torsional modulus. It was determined in a comparative experiment that a direct relationship exists in these systems between modulus and level of crystallinity.

Octadecyl acrylate copolymer had the highest relative rate of crystallization over the range investigated (−20 to 100 °C); the crystallization rate of this copolymer had a maximum value at ∼ 70 °C. The butylacrylate copolymer crystallized most slowly; its rate reached a maximum at $\sim 80\,^\circ$C. The behavior of octyl acrylate copolymer was intermediate. These results correlate with the hypothesis that the difference between T_M and T_G governs the crystallization kinetics. According to the data of Jordan et al.,[32] all of these copolymers have approximately the same T_M but T_G drops substantially from the butyl ($-8\,^\circ$C) to the octadecyl ($-18\,^\circ$C) acrylate copolymer.

The up-swing in crystallinity at very low temperatures which was observed in the octadecyl acrylate copolymer is probably associated with side chain crystallization. A similar effect was reported for N-octadecyl acrylamide copolymers.[33] The crystalline phase formed by side chain crystallization has not been investigated further.

REFERENCES

1. R. D. Lowry and R. M. Wiley, unpublished results, The Dow Chemical Company (1941).
2. L. K. Frevel, unpublished results, The Dow Chemical Company (1938).
3. R. C. Reinhardt, *Ind. Eng. Chem.*, **35**, 422 (1943).
4. A. Sharples, *Introduction to Polymer Crystallization*, St. Martin's Press, N.Y. (1966).
5. G. C. Oppenlander, *Science*, **159**, 1311 (1968).
6. A. Peterlin, *Encyc. Polym. Sci. Tech.*, **9**, 204 (1968).
7. K. Okuda, *J. Polym. Sci., A*, **2**, 1749 (1964).
8. D. Kockott, *Kolloid-Z. Z. Polym.*, **206**, 122 (1965).
9. A. Bailey and D. H. Everett, *J. Polym. Sci.*, A-2, **7**, 87 (1969).
10. P. J. Flory, *Trans. Faraday Soc.*, **51**, 848 (1955).
11. D. Kockott, *Kolloid-Z. Z. Polym.*, **214**, 31 (1967).
12. V. P. Lebedev, N. A. Okladnov and M. N. Shlykova, *Polym. Sci. USSR*, **9**, 553 (1967).
13. R. A. McDonald, Thermal Research Laboratory, The Dow Chemical Company, unpublished results.
14. I. R. Harrison and E. Baer, *J. Colloid Interface Sci.*, **31**, 176 (1969).
15. R. A. Wessling, unpublished results.
16. R. M. Wiley, The Dow Chemical Company, unpublished results.
17. R. W. Eykamp, A. M. Schneider and E. W. Merrill, *J. Polym. Sci. A-2*, **4**, 1025 (1966).
18. R. A. Wessling, *J. Appl. Polym. Sci.*, **14**, 1531 (1970).
19. D. N. Bort and S. A. Arzhakov, *Vest. Akad. Nauk SSSR*, **40**, 79 (1970).
20. R. A. Wessling, J. H. Oswald and I. R. Harrison, *J. Polym. Sci. Phys.*, **11**, 875 (1973).

21. F. Khoury and J. D. Barnes, *J. Res. Natl. Bur. Stand. (U.S.)*, **76A**, 225 (1972).

22. R. A. Wessling, D. R. Carter and D. L. Ahr, *J. Appl. Polym. Sci.*, **17**, 737 (1973).

23. A. F. Burmester and R. A. Wessling, *Bull. A. Phys. Soc.*, **18**, 317 (1973).

24. R. A. Wessling and E. B. Bradford, The Dow Chemical Company, unpublished results.

25. G. Schuur, *Some Aspects of the Crystallization of High Polymers*, Rubber-Stiching, Delft (1955).

26. E. F. Gurnee, The Dow Chemical Company, unpublished results.

27. K. Okuda, R. Kakabe, K. Watanabe, M. Sugita, T. Hotta and M. Asahina, *Kogyo Kag. Zasshi*, **73**, 1398 (1970).

28. H. D. Keith, in Fox, Labes and Weissburger (eds.), *Physics and Chemistry of the Organic Solid State*, Vol. 1, Interscience, N.Y., p. 462 (1963).

29. R. F. Boyer, R. S. Spencer and R. M. Wiley, *J. Polym. Sci.*, **1**, 249 (1946).

30. R. F. Boyer and R. S. Spencer, *J. Appl. Phys.*, **15**, 398 (1944).

31. G. R. Riser and L. P. Witnauer, *SPE Trans.*, **2**, 7 (1962).

32. E. F. Jordan, W. E. Palm, K. P. Witnauer and W. S. Bort, *Ind. Eng. Chem.*, **49**, 1695 (1957).

33. E. F. Jordan, G. R. Riser, B. Artynayshym, W. E. Parker, J. W. Pensabene and A. N. Wrigley, *J. Appl. Polym. Sci.*, **13**, 1777 (1969).

CHAPTER 7

Interaction of PVDC with other Molecules

7.1 SOLUBILITY

7.1.1 Effect of solvent structure

PVDC is generally categorized as an insoluble, intractable material and is impervious to most reagents. Saran copolymers show the same characteristics though to a lesser degree. This creates problems for those interested in studying Saran. But in a more positive vein, it leads to some very useful properties in commercial applications, namely high solvent resistance and low gas permeability. These characteristics result from the crystallinity of the material and the way it interacts with other molecules.

Highly crystalline polymers, like PVDC, tend to be insoluble except at temperatures approaching or above their melting points.[1] The copolymers, because of their lower crystallinity are usually more soluble.

Most polar polymers tend to be insoluble except in specific solvents. For this reason, even amorphous Saran copolymers are difficult to dissolve. But, in general, crystallinity is a more important factor than polarity. PVDC, for example, dissolves in a wide variety of solvents at temperatures above 130 °C (its melting point is 200 °C).[2] But, at lower temperatures it will only dissolve in certain classes of polar aprotic solvents. A list of both good normal solvents and specific solvents is given in Table 7.1.

Early interest in solvents for Saran was generated by a desire to develop lacquer coatings systems. This was accomplished by a combination of copolymerization[3] and selection of solvent.[4]

Both specific solvents like cyclohexanone and solvents of general

TABLE 7.1

Solvents for Polyvinylidene Chloride (Ref. 2)

Nonpolar solvents	T_S °C[a]
1,3-dibromopropane	126
Bromobenzene	129
α-chloronaphthalene	134
2-methylnaphthalene	134
o-dichlorobenzene	135

Polar aprotic solvents	T_S °C
Hexamethylphosphoramide	− 7.2
Tetramethylene sulfoxide	28
N-acetyl piperidene	34
N-methyl pyrrolidone	42
N-formyl hexamethyleneimine	44
Trimethylene sulfide	74
N-n-butyl pyrrolidone	75
Isopropyl sulfoxide	79
N-formyl piperidine	80
N-acetyl pyrrolidine	86
Tetrahydrothiophene	87
N,N-dimethyl acetamide	87
Cyclo-octanone	90
Cycloheptanone	96
n-butyl sulfoxide	98

[a]   Temperature at which a 1% mixture of polymer in solvent becomes
homogeneous,

utility like chlorinated aromatic compounds were identified in early
studies at Dow. A variety of other solvents which are claimed to be
more effective have since been patented.[5-9] These include sulfones,
sulfonamides and phosphonamides as well as the classes of solvents
listed in table 7.1. The same solvents are also effective in dissolving
many other polar polymers such as poly(acrylonitrile) and poly-
(vinyl chloride).[10]

These solvents are classified as polar aprotic solvents.[11] They are
Lewis bases and have high dielectric constants, high dipole moments
and relatively high boiling points. Polar aprotic solvents are good

hydrogen bond acceptors and are, therefore, often water soluble and hydroscopic. They also interact strongly with proton donors like phenol[13] and many chlorinated hydrocarbons.[12,14]

Because of its high crystallinity, the solubility of PVDC can be analyzed in a more quantitative fashion than is the case for poorly crystalline polymers like PVC. The solubility of highly crystalline polymers is described by Flory's theory of melting point depression for polymer solvent mixtures.[15] The melting point depression is given by Equation (7.1).

$$\frac{1}{T_M} - \frac{1}{T_M^\circ} = \frac{RV_u}{\Delta H_u V_1}(v_1 - \chi_1 v_1^2). \tag{7.1}$$

Equation (7.1) states that under equilibrium conditions, the observed solution temperature should depend only on the interaction parameter, χ_1 and on molar volume, V_1. A solvent with small V_1 may be able to dissolve a crystalline polymer even when the polymer-solvent interaction is unfavorable. Therefore, a simple comparison of T_M values may not bring our the relative effectiveness of solvents for the polymer unless they have similar molar volumes.

This concept can be shown by expressing the interaction parameter in the form

$$\chi_1 = \chi_s + \frac{BV_1}{RT_M} \tag{7.2}$$

and rearranging Equation (7.1) to:

$$\frac{1}{T_M} = \frac{1}{T_M^\circ[1 + BV_u v_1^2/\Delta H_u]} + \frac{RV_u(v_1 - \chi_s v_1^2)/\Delta H_u}{[1 + BV_u v_1^2/\Delta H_u]V_1} \tag{7.3}$$

This reduces to a linear equation if the mixture is ideal.

Equation (7.3) suggests that the true affinity of a solvent for the polymer can be illustrated on a $1/T_M$ vs $1/V_1$ plot. This method was used in the solubility study mentioned earlier to group solvents into various classes depending on how they interact with PVDC. Most of the very strong solvents fall in the region where $\chi_1 < 0$. This indicates that they mix exothermically with PVDC as might be expected for a Lewis acid-base interaction.

7.1.2   Effect of polymer history on solubility

The data in Table 7.1 characterize the solubility of "as polymerized" polyvinylidene chloride. It is known that the polymer is not in its most crystalline state. Therefore, recrystallization or annealing will raise solution temperatures. But the relative order should still hold.

As pointed out earlier, annealing of PVDC is complicated since the polymer starts eliminating HCl above $\sim 130\,^{\circ}\mathrm{C}$ (see Chapter 9). Annealing below this temperature does not significantly affect T_M. Above $130\,^{\circ}\mathrm{C}$, the accompanying degradation actually reduces crystallinity. Crystallinity can be increased, however, by annealing for short times in the range of $130\text{-}160\,^{\circ}\mathrm{C}$ since the amount of degradation experienced is insignificant. Polymers annealed in this way have slightly higher solution temperatures.

The interpretation of solubility data in terms of Equation (7.1) is valid only for systems in thermodynamic equilibrium. This condition cannot be achieved experimentally. But the hypothetical solution temperature of a perfectly crystalline specimen can be obtained by extrapolation.[16] This is illustrated by the plot of solution temperature vs crystallization temperature in tetramethylene sulfide as shown in Figure 7.1.[17] The intersection of the extrapolated experimental curve with the equilibrium curve gives the true thermodynamic solution temperature, T_M. This quantity is independent of polymer history and should really be used to characterize and compare solvents.

7.1.3   Solubility parameters

The solubility of amorphous polymers is normally correlated with cohesive energy densities or solubility parameters. Burrell has reported a value of 12.2 for the solubility parameter of PVDC.[18] This is much higher than a value estimated from solubility studies in nonpolar solvents $(10.1 \pm 0.3)^2$ or calculated from Small's relationship,[19]

$$\delta = \frac{d\,\Sigma G_i}{M}\,, \tag{7.4}$$

where δ is density, G_i is a molar attraction constant and M, the molecular weight. Using the value of $1.7754\ \mathrm{g/c^3}$ (Ref. 20) for the

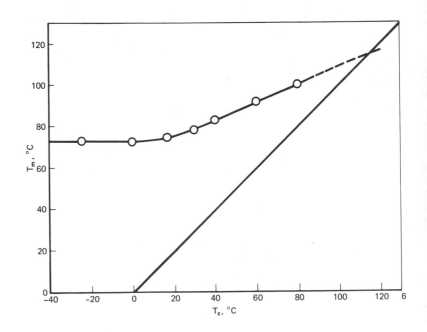

FIGURE 7.1   Effect of crystallization temperature on solution temperature
for PVDC in tetramethylene sulfide (0.25 g/15 ml).

density of amorphous PVDC, the calculated value of δ is 10.2. The
low values are more likely to be correct. In any case, the use of the
solubility parameter scheme for polar crystalline polymers like
PVDC is of limited value.

Copolymers with a high enough vinylidene chloride content to
be quite crystalline, behave much like PVDC. They are more
soluble, however, because of their lower melting points. The solubility of amorphous copolymers is much higher. The selection of
solvents in either case, varies somewhat with the type of copolymer. Some of the more common types are listed in Table 7.2.
Solvents that dissolve PVDC will also dissolve the copolymers, but
at lower temperatures. The solubility parameters for copolymers

depend on composition. Measured values range from 9.5 to 14.7.[21]

TABLE 7.2

Common Solvents for Saran Copolymers

Solvent	Copolymer type	Range $^{\circ}$C
THF	All	<60
MEK	Low crystallinity polymers	<80
1,4-dioxane	All	50-100
Cyclohexanone	All	50-100
Cyclopentanone	All	50-100
Ethyl acetate	Low crystallinity polymers	<80
Chlorobenzene	All	100-130
Dichlorobenzene	All	100-140
DMF	High AN	<100

7.1.4   Solubility in solvent mixtures

PVDC also dissolves readily in certain solvent mixtures where one component is a sulfoxide or amide of the type described above and the cosolvents are less polar and usually have cyclic structures. They include both aliphatic and aromatic hydrocarbons, ethers, sulfides, and ketones. Acidic or hydrogen bonding solvents have an opposite effect, rendering the polar aprotic component less effective.

This behavior can be qualitatively predicted from the Flory Huggins theory for a 3-component (polymer-solvent A-solvent B) mixture.[15] A theoretical analysis[22] predicts the existence of ideal mixtures (T_M vs solvent composition, (v_1), plots are linear); favorable mixtures that are better than ideal, $d^2 T_M/dv_1{}^2 > 0$; and mixtures that are poorer than ideal, $d^2 T_M/dv_1{}^2 < 0$. It also predicts that certain cosolvents should show a maximum or minimum in the T_M vs v_1 plot. Examples that appear to fit all of these categories have been observed experimentally.[22]

Assuming that systems showing linear T_M vs v_1 plots are normal mixtures, the linearity requires that $\delta_1 = \delta_2$. This does not appear to be the case. Other examples of normal mixtures are better than ideal but solubility is enhanced only where $\delta_1 < \delta_p < \delta_2$. Mixtures of nonpolar solvents do not depress T_M much below its value in the better component.

When one or both of the solvent components is a polar aprotic solvent like those described earlier, melting point depressions can be quite large or much less than expected depending on the nature of the cosolvent. These mixtures are described as "favorable" and "unfavorable", respectively.

A favorable mixture was defined as one in which addition of a cosolvent to the base solvent either lowers the solution temperature or does not significantly increase it. (The base solvent being the better of the two.) The behavior of "favorable" mixtures is shown in Figure 7.2.

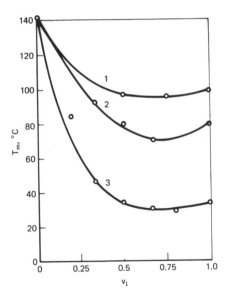

FIGURE 7.2 Solution temperatures for PVDC in favorable mixtures (1% polymer): (1) tetrahydronaphthalene-isopropyl sulfoxide; (3) tetrahydronaphthalene-tetramethylene sulfoxide.

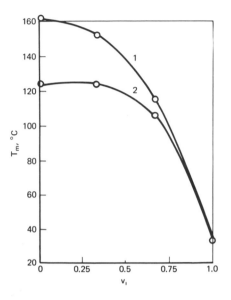

FIGURE 7.3 Solution temperatures for PVDC in unfavorable mixtures (1% polymer): (1) benzyl alcohol-tetramethylene sulfoxide; (2) bromoform-tetramethylene sulfoxide.

The "unfavorable" solvent mixtures were defined as those where the addition of cosolvent raised the solution temperature sharply. In these cases, the mixture is poorer than ideal. This behavior was observed primarily in mixtures of dipolar aprotic solvents with H-bonding solvents as shown in Figure 7.3.

Changes of solvent activity with composition can also be observed by viscosity measurements.[23] Plots of $[\eta]$ vs composition of a favorable mixture at two temperatures are shown in Figure 7.4. Since $[\eta]$ is related to the size of the polymer coil in solution, it increases with solvent power. The maximum correlates with the most favorable composition found from T_M studies.

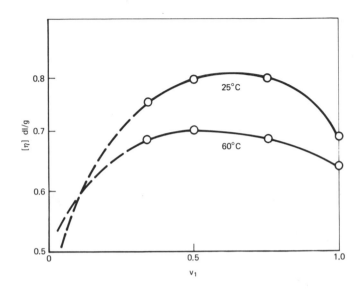

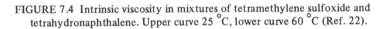

FIGURE 7.4 Intrinsic viscosity in mixtures of tetramethylene sulfoxide and tetrahydronaphthalene. Upper curve 25 °C, lower curve 60 °C (Ref. 22).

7.2 SOLUTION PROPERTIES

Neither 1% solution temperature nor solubility parameter is a very adequate way of evaluating the interaction parameter. A better technique is to analyze the phase diagram by Flory's method.[15] Typical phase diagrams for PVDC[24] in a good solvent, TMSO and in a nonpolar solvent, tetrahydronaphthalene (THN), are shown in Figure 7.5. The derived parameters are listed in Table 7.3. The ΔH_u value obtained in THN is in agreement with other work (see Chapter 6 for discussion). A high value was obtained in TMSO by using unannealed samples. The entropy change ΔS_u is typical of values observed on polymers with a moderately stiff chain. The negative value of B found for TMSO and the large positive value in THN, agree with the predictions from the solubility study.

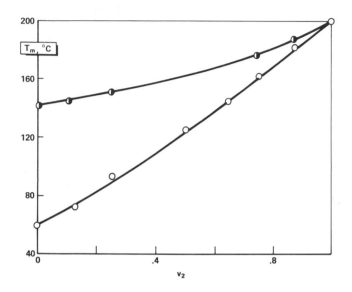

FIGURE 7.5 Phase diagrams for mixtures of PVDC with tetrahydronaphthalene (upper curve), and tetramethylene sulfoxide (lower curve).

TABLE 7.3
Parameters derived from Analysis of Phase Diagrams

Solvent	ΔH_u kcal/mole	ΔS_u cal/mole/deg	B cal/cm^3
Tetramethylene sulfoxide	1.69	3.6	−2.53
Tetrahydronaphthalene	1.57	3.3	+2.13

The study of PVDC in poor solvents where liquid-liquid phase separation takes place is ruled out by problems with polymer stability. Though liquid-liquid phase separation can be observed in borderline solvents like 1-octanol at ∼ 160 °C, it is not possible to reliably

determine the phase diagrams. In solvents where PVDC dissolves
at a low enough temperature to form stable solutions, only liquid-
crystalline phase separation is observed.

Problems with polymer degradation also limit intrinsic viscosity
and light scattering measurements to good solvents. o-Dichlorobenzene was used extensively as a solvent for dilute solution studies
by early Dow workers. Its major drawback is that measurements
must be carried out at $> 120\,°C$ to keep the polymer in solution.
It was common practice to correlate molecular weights with a 2%
solution viscosity at $140\,°C$. The exact correlation between this
number and molecular weight was never reported.

Dilute solution studies in strong solvents were not described until
the very recent work of Matsuo and Stockmayer.[25] They obtained
intrinsic viscosity data at $25\,°C$ and used light scattering to
measure the molecular weights of unfractionated homopolymer
samples. Mark-Houwink relationships were derived for PVDC in
three solvents as shown below:

$$[\eta] = 1.31 \times 10^{-4} \bar{M}_v^{0.69} \quad (N\text{-methylpyrrolidone}) \qquad 7.5$$
$$[\eta] = 1.39 \times 10^{-4} \bar{M}_v^{0.69} \quad (\text{Tetramethylenesulfoxide}) \qquad 7.6$$
$$[\eta] = 2.58 \times 10^{-4} \bar{M}_v^{0.65} \quad (\text{Hexamethylphosphoramide}) \; 7.7$$

The change in viscosity with solvent follows the order predicted
by solution temperature measurements.

The unperturbed dimensions of the PVDC chain were estimated
by the Stockmayer-Fixman technique.[26] K_θ was obtained from a
plot of $[\eta]/M^{1/2}$ vs $M^{1/2}$. The derived quantities, $\langle r^2 \rangle_0/M$ the
unperturbed dimension, and C_∞, the characteristic ratio are compared in Table 7.4 to values obtained for related polymers.

The comparison shows that the PVDC chain is more expanded.

Matsuo and Stockmayer concluded, on the basis of conformational energy calculations, that the larger dimensions of the PVDC
chain were a result of electrostatic dipole interactions.

Many of the copolymers of VDC can be dissolved in solvents such
as THF conventionally used for dilute solution studies. Molecular
weights of copolymers have been measured in THF, MEK, and o-
dichlorobenzene among others. Techniques include viscosity, osmo-
metry, light scattering, gel permation chromatography and NMR

TABLE 7.4

Comparison of Chain Dimensions

Polymer	$\langle r^2 \rangle_0/M \times 10^{18}$ cm^2 mol/g	C_∞	Reference
Polyethylene	30.5-32.3	6.6-6.8	27
Polyisobutylene	10.7	6.6	27
PVDC	39 ± 4	8 ± 1	25

measurements.

Wallach[28] has carried out a classical diulte solution study on a Saran F resin (F-216). This is a commercial polymer containing 9 wt. % acrylonitrile. The whole polymer was fractionated and both light scattering molecular weights and intrinsic viscosities were measured on the fractions. The Mark-Houwink relationship is

$$[\eta] = 1.06 \times 10^{-4} \bar{M}_w^{0.72} . \qquad (7.8)$$

The unperturbed dimensions of the polymer chain calculated from a plot of $[\eta]/\bar{M}_w^{1/2}$ vs $\bar{M}_w^{1/2}$, are $\langle r^2 \rangle_0/M = 47 \times 10^{-18}$ cm^2 mol/g and characteristic ratio $C_\infty = 8.8$. These values suggest that the copolymer coil may be somewhat more expanded.

The dilute solution properties of several acrylate copolymers have also been studied.[29] The authors carried out phase equilibrium studies to get θ temperatures and then measured intrinsic viscosity-molecular weight relationships at the θ temperatures. Molecular weights were determined by light scattering.

The phase diagrams for the octyl acrylate copolymer are shown in Figure 7.6. The θ temperature was obtained by a plot of $1/T_c$ vs $M^{-1/2}$. The dimensions of the polymer coil at the θ temperature were measured. The results are collected in Table 7.5. This study shows that the unperturbed dimensions of these copolymers are similar to those of the homopolymer. The length of the acrylate ester side chain had little effect on properties.

7.3 PERMEABILITY

Vinylidene chloride polymers are very impermeable to a wide variety

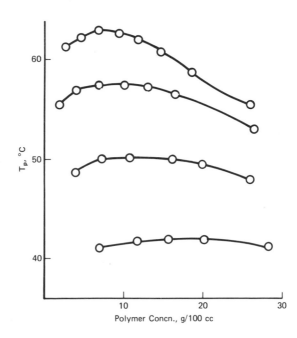

FIGURE 7.6 Phase separation temperature, T_p, of four fractions of poly-(vinylidene chloride co n-octyl acrylate) as a function of polymer concentrations (Ref. 29).

of gases and liquids. A comparison of permeability of a commercial Saran film (VDC/VC copolymer) with that of various other polymer is shown in Table 7.6.[30]

Saran is a substantially better barrier to all the gases tested than any of the other materials. This can be shown to be a consequence of its structure and the way it interacts with gas molecules.

For a semi-crystalline polymer, permeability depends on three parameters as shown in Equation (7.9).[31]

$$P = D_a S_a (1 - \varphi)^2, \tag{7.9}$$

where D_a is the diffusion coefficient for the pentrant in the

TABLE 7.5

Chain Dimensions of Saran Copolymers (Ref. 29)

Copolymer, Mole %	Theta temp.	$K_\theta{}^a$	$(r_0{}^2/M_w)^{1/2}$	$(r_0{}^2/r_0{}^2)^{1/2}{}_{MEK}$
Ethyl acrylate	49.6 °C			
Ethyl acrylate, 14.9%	Ethyl aceto-acetate 44.0 °C	6.64×10^{-4}	6.81×10^{-9}	2.18
n-butyl acrylate, 16.7%	Benzyl alcohol 56.8 °C	7.30×10^{-4}	7.03×10^{-9}	2.31
n-hexyl acrylate, 14.5%	Benzyl alcohol 77.9 °C	5.50×10^{-4}	6.04×10^{-9}	2.14
n-octyl acrylate, 15.6%	Benzyl alcohol	5.87×10^{-4}	6.53×10^{-9}	2.23

[a] $[\eta]_\theta = K_\theta M_w{}^{1/2}$.

TABLE 7.6

Typical Gas Transmission Rates of Plastic Films at 23 °C[a] (Ref. 30)

Film type	Permeability c^3/m^2-24 hr-1 atm.		
	O_2	N_2	CO_2
Cellulose acetate	350	1500	7800
Methylcellulose	1300	450	6800
Polyethylene 0.917 density	2700		
Polyethylene 0.960 density	1600	440	
Polyethylene terephthalate	50	8.4	240
Polystyrene	4500	640	11000
Polyvinyl chloride, plastized	190-3100	53-810	430-19,000
Polyvinyl chloride, rigid	120	20	320
Rubber hydrochloride	390	62	1100
Saran	16	2.5	50
Styrene-acrylonitrile copolymer	900	120	2800

[a] ASTM method D1434-56T.

amorphous polymer, S_a is the solubility coefficient of the penetrant in the amorphous polymer, and φ is the volume fraction of the crystalline phase.

This model is based on the assumption that the crystalline phase is impenetrable.

According to Equation (7.9) permeability can be a function of morphology but we want to look mainly at the factors affected by chemical composition. The best way to do this is to compare the permeability of several polymers to a common penetrant. Lasoski[32,33] has published some data of this type with the objective of showing how symmetry affects permeability of a polymer to moisture. His data, slightly modified and supplemented are shown in Table 7.7. The importance of symmetry is quite evident in comparing permeability in the amorphous polymers.

TABLE 7.7

Comparison of the Permabilities of Various Polymers to Water Vapor at 39.5 $^{\circ}$C

Polymer	Density[a]		Permeability[a]	
	Amorphous	Crystalline	Amorphous	Crystalline
Polyethylene	0.85 g/c^3	1.00	200-220	10-40
Polypropylene	0.85	0.94	420	–
Polyisobutylene	0.915	0.94	90	–
Poly(vinyl chloride)	1.41	1.52	~300	90-115
Poly(vinylidene chloride)	1.77	1.96	~ 30	4-6

[a] Units = g/100 in^2/h/ml.

Polymer composition influences permeability in the amorphous polymers through the constants D_a and S_a. Considering first solubility, one can make a correlation between permeability and the cohesive energy densities of polymer and penetrant. On this basis, we can predict that polyethylene will be a good barrier to water but a poor barrier to pentane. The opposite should be true for polyvinyl alcohol. PVDC with a solubility parameter in the middle of the scale should be a good barrier to either. Solubility differences might account for the results in Table 7.6 but cannot explain the results in Table 7.7.

The large differences in permeability of amorphous polymers of similar cohesive energy density must be related to differences in D_a. The data imply that D_a is increased by asymmetry in the polymer structure. If D_a is related to T_G, then the ability of a penetrant to diffuse through a polymer matrix can be put into the same frame of reference as the melting and glass transitions by using Kanig's model[34] discussed in Chapter 8. The vibrational volume increases with temperature and is roughly the same for all polymers. Hole concentration, however, is constant below T_G and increases with T-T_G. Both T_G and the number of holes present below T_G are affected by polymer structure.

Returning now to the effect of symmetry, it seems reasonable that the number of holes at T_G should be greatest for a polymer with a bulky and irregular structure. The permeability of this type of polymer should be relatively high, other factors being constant. A polymer with a smooth symmetrical structure should have a much lower free space and correspondingly a lower D_a.

Direct measurements of the barrier properties of PVDC have not been reported. By comparison with copolymers, however, we would expect very low permeability to most gases and vapors. The permeability of a Saran film to a variety of different gases is shown in Table 7.8.

Based on the above results with a VDC/VC copolymer, one would predict that PVDC itself has broad spectrum barrier properties. Other comonomers of similar cohesive energy density to VDC would also yield copolymers which are good barriers to a variety of gases. Methyl acrylate copolymers, for example, might be expected to have similar characteristics though they would be more sensitive to H-bonding penetrants or polar gases of acidic character. Neither PVDC nor any of its copolymers can be expected to be a good barrier to strong Lewis bases.

7.4 SURFACE PROPERTIES

Zisman and coworkers have studied the surface properties of numerous polymers including PVDC and various copolymers.[36-38] They measured both the coefficient of friction[38] and the critical

TABLE 7.8
Permeability of a Saran B film to Various Gases[a] (Ref. 35)

Gas	Temp $^\circ$C	Act. energy E_p	$P \times 10^{10}$
N_2	30	16.8	0.00094
O_2	30	15.9	0.00053
CO_2	30	12.3	0.03
He	34	–	0.31
H_2O	25	11.0	0.5
H_2S	30	17.8	0.03

[a]   Units: cm^3 (STP) $cm/cm^2 sec.$ cmHg.

TABLE 7.9
Surface Properties of PVDC and Saran Copolymers at 25 $^\circ$C

Polymer	Critical contact angle, $^\circ$	Contact angle			Coefficient of friction	
		H_2O	Methylene iodide	Hexa-decane	Static	Kinetic
PVDC	40	83	30	0	0.50	0.45
Saran F (20% AN)	38-44	81	32	0	0.80	0.65
Saran B-1500	39-40	80	31	0	0.49	0.46

surface tension for wetting.[33] These data are listed in Table 7.9.

The surface properties of Saran are greatly altered by the incorporation of surface active additives in the polymer. When heated above some initial temperature, these species migrate to the surface of the polymer. This phenomenon has been studied in detail by Owens.[39-41] Effective additives include fluorinated compounds and various polar waxes like stearamide. They tend to reduce the coefficient of friction and make the polymer surface slippery.

Owens[42] has also analyzed the components of the surface free energy of polymers. He reports a value of 45 ergs/cm^2 for PVDC. This is some what larger than the critical surface tension for wetting, a difference observed in most other polymers also.

REFERENCES

1. H. Morawetz, *Macromolecules in Solution*, 2nd edition, Interscience, N.Y., Chapter II (1975).
2. R. A. Wessling, *J. Appl. Polym. Sci.*, **14**, 1531 (1970).
3. R. M. Wiley, U.S. 2,160,945 (1939); U.S. 2,235,782 (1941), to The Dow Chemical Company.
4. R. C. Reinhardt and J. H. Reilly, U.S. 2,249,915; 2,249,916; 2,249,917 (1941), to the Dow Chemical Company.
5. R. C. Houtz, U.S. 2,460,758; 2,460,759 (1949), to du Pont.
6. G. F. D'Alelio, U.S. 2,531,406 (1950), to Industrial Rayon Corp.
7. G. E. Ham, U.S. 2,587,464 (1952) to Eastman Kodak.
8. H. W. Coover, Jr. and J. B. Dickey, U.S. 2,742,444 (1956).
9. D. A. Baggett and H. H. McClure, U.S. 2,886,547 (1959), to the Dow Chemical Company.
10. G. E. Ham, *Ind. Eng. Chem.*, **46**, 390 (1954).
11. A. J. Parker, *Quart. Rev. (London)*, **16**, 163 (1962).
12. E. Halpern, J. Houck, H. Finegold and J. Goldenson, *J. Am. Chem. Soc.*, **77**, 4472 (1955).
13. D. P. Eyman and R. S. Drago, *J. Am. Chem. Soc.*, **88**, 1617 (1966).
14. H. Normant, *Angew. Chem. (Int. Eng. Ed.)*, **6**, 1046 (1968).
15. P. J. Flory, *Principles of Polymer Chemistry*, Cornell Univ. Press, Ithaca, N.Y., Chapter 13 (1953).
16. J. F. Kenny and V. F. Holland, *J. Polym. Sci. A-1*, **4**, 699 (1966).
17. R. A. Wessling, D. R. Carter and D. L. Ahr, *J. Appl. Polym. Sci.*, **17**, 737 (1973).
18. H. Burrell, Official Digest, p. 726 (1955).
19. P. A. Small, *J. Appl. Chem.*, **3**, 71 (1953).
20. R. W. Eykamp, A. M. Schneider and E. W. Merrill, *J. Polym. Sci. A*, **4**, 1025 (1966).
21. H. Burrell and B. Immergut, in J. Brandup and E. H. Immergut (eds.), *Polymer Handbook*, Interscience, N.Y., pp. IV-360 (1966).
22. R. A. Wessling, *J. Appl. Polym. Sci.*, **14**, 2263 (1970).
23. R. A. Wessling, *J. Appl. Polym. Sci.*, **17**, 503 (1973).
24. R. A. Wessling, unpublished results.
25. K. Matsuo and W. H. Stockmayer, *Macromolecules*, **8**, 660 (1975).
26. W. H. Stockmayer and M. Fixman, *J. Polym. Sci.*, **C1**, 137 (1963).
27. P. J. Flory, *Statistical Mechanics of Chain Molecules*, Interscience, N.Y., Chapter II, Table I (1969).
28. M. L. Wallach, ACS, *Polymer Div. Preprints*, **10**, 1248 (1969).
29. M. Asahina, M. Sato and T. Kobayashi, *Bull. Chem. Soc. Japan*, **35**, 630 (1962).
30. W. E. Brown and W. J. Sauber, *Modern Plastics*, p. 110 (1959).
31. P. Meares, *Polymers, Structure and Bulk Properties*, D. Van Nostrand, N.Y., Chapter 12 (1965).

32.　S. W. Lasoski, *J. Appl. Polym. Sci.*, **4**, 118 (1960).
33.　S. W. Lasoski and W. H. Cobbs, *J. Polym. Sci.*, **36**, 21 (1959).
34.　G. Kanig, *Kolloid-Z. Z. Polym.*, **190**, 1 (1963).
35.　H. Yasuda, in J. Brandup and E. H. Immergut (eds.), *Polymer Handbook*, Interscience, N.Y., p. V-13 (1966).
36.　A. H. Ellison and W. A. Zisman, *J. Phys. Chem.*, **58**, 260 (1954).
37.　W. A. Zisman, *Adv. Chem. Ser.*, **43**, 1 (1964).
38.　R. C. Bowers, N. L. Jarvis and W. A. Zisman, *Ind. Eng. Chem. Prod. R & D*, **4**, 86 (1965).
39.　D. K. Owens, *J. Appl. Polym. Sci.*, **8**, 1465 (1964).
40.　D. K. Owens, *J. Appl. Polym. Sci.*, **14**, 185 (1970).
41.　D. K. Owens, in O. J. Sweeting (ed.), *The Science and Technology of Polymer Films*, Vol. I, Interscience, N.Y., Chapter 9 (1968).
42.　D. K. Owens, *J. Appl. Polym. Sci.*, **13**, 1741 (1969).

CHAPTER 8

Transitions and Mechanical Behavior

8.1 TRANSITIONS IN PVDC

The properties of a crystalline polymer like PVDC change drastic-ally when the material is heated or cooled. Major changes take place at two well-defined temperatures: the melting point (T_M); and the glass transition temperature (T_G). The polymer is transformed from a tough leathery material to a rigid brittle solid when cooled through T_G; when heated above T_M, it changes to a soft rubber or a visco-elastic fluid. The location of these temperatures with respect to the use temperature range determines the character of the material.[1]

The melting and glass transitions can be detected by various methods. Differential thermal analysis (DTA) is an excellent method for measuring the melting point of PVDC,[2] but other fast methods such as differential scanning calorimetry[3] and hot stage micro-scopy[4] can also be used. Degradation of the polymer interferes with slow methods of measuring T_M; therefore, the classical techniques of dilatometry and calorimetry have been used mainly to investigate changes in PVDC at low temperatures.

The heat capacity of PVDC from 15-300°K has been measured, and its thermodynamic properties calculated from these data.[5] Only one transition, at approximately 260 °K was observed over this temperature range.

This has been identified by Boyer and Spencer[6-8] as the glass transition temperature. In their studies dilatometry was used to measure coefficients of expansion above and below T_G. Com-parison of data on crystalline and amorphous specimens show that T_G is not affected by level of crystallinity. The coefficient of expansion above T_G, however, is smaller in crystalline polymers.

Other methods of detecting transitions, including NMR, IR, dielectric measurements and dynamic mechanical measurements, can be applied up to temperatures approaching T_M. The dynamic methods show the presence of a transition between T_M and T_G. The transition temperatures and related properties of PVDC are listed in Table 8.1.

TABLE 8.1

Transition Temperatures and Related Properties of PVDC

Property	Value	Reference
T_M	198-205 °C	2
T_α	80 °C	9
T_β	12 °C	9
T_G	−18 °C	6
d_a, 25 °C	1.775 g/c^3	10
d_c 25 °C	1.90 g/c^3	2
d_c, (unit cell)	1.96 g/c^3	11
α_L (above −17 °C)	5.7 × 10^{-4}	12
α_c (above −17 °C)	4.8 × 10^{-4}	6
α_G (below −17 °C)	1.2 × 10^{-4}	12
n_D^{25}	1.63	13

A DTA trace obtained on PVDC in the "as polymerized" form does not show a well-defined break at T_G.[2] This is due to the low fraction of amorphous material in such samples and the relative insensitivity of the method to small changes in heat capacity. To get a well-defined T_G requires a sample that has been first melted and quenched before heating through T_G. Values of T_G obtained in this way tend to be approximately 10 °C higher than dilatometric values, probably due to the higher heating rates, as well as the quenched state of the sample.

Dynamic mechanical measurements can also be used to detect T_G. The transition occurs at the temperature of the lower lying peak in a plot of loss factor vs temperature (Figure 8.1). The temperature at which it is observed depends on the frequency of the test.

Oscillation at ∼ 1 hertz in a torsion pendulum for example

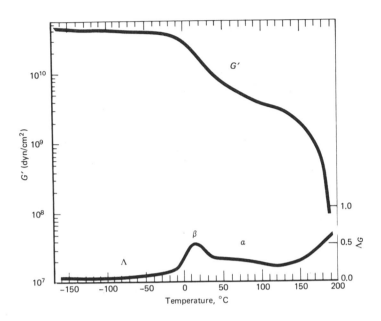

FIGURE 8.1 Temperature dependence of the shear modulus and logarithmic decrement at low frequencies for PVDC (Ref. 9).

results in maxima at ~10 °C. This transition, sometimes referred to as the beta transition (T_β), also correlates with the brittle point (T_B) of the polymer in an impact test. As mentioned above, the dynamic methods show a transition between T_M and T_G. The peak at 80 °C observed in the mechanical loss plot has been identified by McCrum et al.[14] as an alpha transition (T_α). This type of transition is characteristic of semi-crystalline polymers. It appears to be associated with chain motion in the crystal or motion in the chain fold area of the crystal surface. The alpha transition is very prominent in the NMR spectra of PVDC single crystals.[15]

There may be a correlation between T_α and the softening temperature of the polymer. PVDC powders begin to deform in a thermomechanical test slightly above 100 °C.[16] This is still

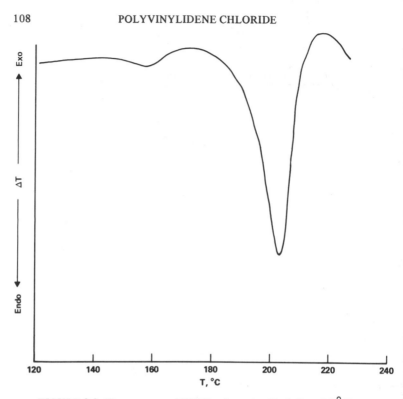

FIGURE 8.2 Thermogram of PVDC polymerized in bulk at 25 °C.

substantially below the melting point. Apparently, the material
softens at T_α to a more rubbery state but does not flow. The heat
distortion temperature (T_D) defined by a tensile test usually correlates with the onset of melting in crystalline polymers. This is
approximately 185 °C for PVDC.[17]

The DTA trace of an "as polymerized" PVDC powder does not
show T_α either.[2] A typical example obtained on a powder isolated
from a low conversion mass polymerization is shown in Figure 8.2.
The first detectable change takes place at a temperature well above
T_α. The main endotherm which peaks at 200 °C has been shown
to represent the melting point of the polymer. Microscopy indicates
melting over a range from ~195 to 205 °C with the last traces of
birefringence disappearing at the upper limit. The characteristic

x-ray diffraction pattern shown by the crystalline polymer disappears in the same temperature range as further proof that this is a melting process.

DTA traces of "as polymerized" powders characteristically show two endothermic peaks. The origin of the low temperature peak is less certain. No changes in the sample can be detected either visibly or by x-ray diffraction at this temperature. The endothermic nature suggests, however, that it represents a partial melting process. It is not characteristic of the polymer because the size and temperature of the lower peak vary with polymerization conditions. In addition, it can be eliminated by annealing or recrystallizing the specimen. It seems to be associated with material that formed in an amorphous condition during polymerization and subsequently crystallized under less favorable conditions.[2,18]

Recrystallization of PVDC from the melt has to be carried out below 130 °C to prevent degradation. Therefore, the melting point is lower than for "as polymerized" powders. The melting point of a crystalline polymer should increase with crystallization temperature over a certain range and plots of T_M vs T_C can be extrapolated to get the equilibrium melting point,[19] but this has not been done for PVDC. The expected effect has been observed in a commercial Saran B resin.[20] But, even in this case, crystallization close to the expected equilibrium melting point cannot be realized due to degradation. Recrystallization from dilute solution is the only way to get high melting recrystallized polymers.

Solvents and plasticizers reduce both T_M and T_G. The effect on T_M can be predicted accurately from Flory's melting point depression equation if the interaction parameter for the particular system is known. For low levels of plasticizer, however, the reciprocal melting point is roughly linear in plasticizer concentration (see Equation (7.1)).

The effect of plasticizer on T_G is complicated by the polymer crystallinity. Plasticized amorphous polymers normally follow an ideal additivity rule like the Gordon-Taylor equation.[21] But in crystalline polymers, the plasticizer is confined to the amorphous regions. Therefore, its effective concentration is higher than the apparent concentration and increases with % crystallinity. Since T_G is a property of the amorphous phase, it falls with increasing crystallinity while the T_G of the unplasticized polymer is un-changed.[6]

The measured values of the transition temperatures are dependent on the type of test used to detect them. Some typical values of T_G obtained by various methods are listed in Table 8.2. The slow cooling tests like dilatometry give the lowest values. High frequency experiments like dielectric loss measurements or broad line NMR give the highest values. DTA results fall in the middle range. In the latter experiment, tests must be carried out on a heating cycle. Therefore, the exact value is very dependent on sample pretreatment.

The effect of rate on T_G is most apparent in the dielectric loss measurements. The range shown in Table 8.2 is for the loss maxima. The "true" value of T_G obtained from the temperature dependence of the area under the loss peaks, was reported to be $-18\ °C^{22}$ in good agreement with the dilatometric values. The association of dielectric loss peaks with the T_G transition is also consistent with observations on other crystalline polymers.

TABLE 8.2

Glass Transition Temperatures by Various Methods

Method	$T_G, °C$	Reference
Pressure transducer	-20.6	23
Dilatometry	-18 to -15	6, 22
DTA	-11 to -1	2
Torsion pendulum	10 to 15	9
Dielectric loss (10^{-1}-10^5 Hz)	-5.5 to 50	22
NMR	$+40$	15

Since the above dielectric data were obtained on PVDC powders, the values have only qualitative significance. Quantitative data have been reported for PVDC film plasticized with 10% phenyl glycidyl ether.[24] The dielectric loss is due to energy absorption by the carbon chlorine dipoles. The dipole moment of PVDC has not been measured, but these results indicate that it is probably lower than that of PVC. However, the lower energy absorption for PVDC may be due to the fact that the dipoles in the crystalline phase are not free to respond to the varying electric field.

The temperatures at which transitions occur in PVDC are consistent with current concepts of the relation between polymer structure and properties.[1,25] On a molecular level the effect of structure on transitions can be viewed as the result of three factors: Interactions between neighboring groups on the same chain; interactions between groups on different chains; and packing phenomena. The first relates primarily to chain stiffness, the second to cohesive energy; the third is a less well-defined concept relating to free volume in the solid.

There is a large body of data showing that T_G increases with chain stiffness. The flexibility of a polymer chain can be altered either by changing the backbone structure or by introducing substituents on the chain. The strength of interchain forces also affects T_G significantly. Non-polar polymers held together by weak Van der Waals forces have a relatively low T_G. Ionic polymer held together by strong electrostatic forces have a high T_G. Polar polymers containing dipoles or H-bonding sites occupy an intermediate position.

When the polymer has a regular structure and can crystallize, the same forces that raise T_G also raise T_M. The correlation is quite striking for polymers with symmetrical structures where the simple relationship $T_M \cong 2 T_G$ (in $^\circ K$) is followed by many polymers. PVDC fits this correlation reasonably well; $T_M = 1.83 T_G$. Generally, the proportionality constant is between 1.5 and 2.0 for all crystalline polymers.[26]

Application of this rule of thumb requires an accurate knowledge of T_M. In most polymers, however, measured values of T_M are substantially lower than the thermodynamic equiilibrium value. This is true even for polyethylene, the most easily crystallized polymer. The equilibrium T_M of PVDC is not known with certainty but by comparison with other polymers, it is probably about 10 $^\circ C$ higher than the "as polymerized" T_M. The highest value reported in the literature is 210 $^\circ C$; the method of measurement was not reported.[27]

A more subtle aspect of the relationship between structure and transition temperatures is symmetry. Table 8.3 shows the large differences observed in T_G when comparing polymers of similar chemical composition, one made up of symmetrical units and the

TABLE 8.3
Effect of Symmetry of T_G

Symmetrical	T_G, °C	Asymmetrical	T_G, °C
$-CH_2-\underset{\underset{CH_3}{\vert}}{\overset{\overset{CH_3}{\vert}}{C}}-$	−70	$-CH_2-\underset{\underset{H}{\vert}}{\overset{\overset{CH_3}{\vert}}{C}}-$	−20
$-CH_2-\underset{\underset{F}{\vert}}{\overset{\overset{F}{\vert}}{C}}-$	−45	$-CH_2-\underset{\underset{H}{\vert}}{\overset{\overset{F}{\vert}}{C}}-$	+40
$-CH_2-\underset{\underset{Cl}{\vert}}{\overset{\overset{Cl}{\vert}}{C}}-$	−17	$-CH_2-\underset{\underset{H}{\vert}}{\overset{\overset{Cl}{\vert}}{C}}-$	+75

other asymmetrical units.

In the case of VDC and isobutylene (IB), the 1,1-disubstitution might be expected to stiffen the chain as a result of steric interactions between substituents. The fact that T_G is low for these polymers cannot be attributed to chain flexibility in the usual sense. In fact, the less sterically hindered 1,1-dimethyl substituted olefins have a higher T_G than PIB.[14] The cohesive energy densities of these polymers are nearly identical. Therefore, the reason for the difference in T_G must lie elsewhere. Wurstlin[28] has proposed that in the case of the polar polymers, T_G differences are due to reduced dipole interactions in the symmetrical polymers. The carbon-chlorine dipoles in PVDC are internally compensated and hence the net dipole moment per group should be less than in PVC. But this explanation cannot account for the behavior of the methyl substituted polymers where dipole interactions are absent.

Gibbs and DiMarzio[29] have advanced a theoretical explanation based on the assumption that T_G depends on the difference between potential energy minima of rotational isomers. If the difference in conformational energies is large, the polymer chain takes on a preferred conformation. This in effect stiffens the chain and, therefore, leads to a high T_G. If the energy difference between isomers is small (even though the absolute energy might be high), the chain adopts no preferred conformation. In other words, all conformations are equally unfavorable. This gives rise to a more "flexible" chain and, therefore, a lower T_G.

The explanation advanced by Gibbs and DiMarzio is by no means universally accepted.[25] It assumes that chain stiffness is the only factor influencing T_G. Therefore, it cannot account for the high T_G observed in various organic glasses that are not made up of chain molecules. In addition, the ability of polar interactions to raise T_G is given no weight in this treatment.

An equally reasonable explanation for the effect of symmetry on T_G can be deduced from packing considerations. This is implicit in the theoretical treatment advanced by Kanig.[30] The liquid polymer is considered to be a mixture of occupied space and free volume. The free volume is composed of two parts, the oscillation volume (which decreases smoothly with temperature becoming zero at $0\,^\circ K$) and holes. The number of holes also decreases with temperature. But the mobility of the polymer chain depends on the availability of holes. Consequently, as the temperature falls, the number of holes decreases until the polymer chains become frozen into a fixed conformation. Further cooling can no longer reduce the number of holes and the system becomes metastable below T_G. The further volume contraction below T_G is due to loss of oscillation volume.

It seems reasonable to conclude that the number of holes left at T_G is related to the symmetry of the polymer chain. A chain of irregular structure like PVC will require more holes before it can move about. A symmetrical chain with smooth contours like PVDC will be able to move more readily and hence must be cooled to a lower temperature where there are fewer holes before it becomes locked into a single conformation. This picture ties together the melting and glass transitions on a common basis as suggested by the correlation between T_M and T_G mentioned above.

8.2 EFFECT OF COPOLYMER COMPOSITION ON TRANSITIONS

The properties of PVDC are usually modified by copolymerization. Copolymers of high VDC content have lower melting points and usually have higher glass transition temperatures than PVDC itself. As comonomer level is increased, the melting point falls and eventually a point is reached where the polymer is completely amorphous.

The correlation between T_M and T_G does not hold for copolymers. This does not, however, contradict the thesis that the same molecular characteristics influence both transitions. The difference is that T_G is dependent only on relatively short range effects while T_M is very sensitive to long range order. Therefore, a modification in structure that breaks up regularity may have little effect on T_G but will drastically depress T_M. Various comonomers may raise, lower, or leave unchanged the T_G of the copolymer with respect to that of the base polymer. But all will depress T_M more or less by the same amount.

The effect of comonomer content on T_M has been analyzed by Flory[31] who treated the case in which the comonomer unit is completely excluded from the crystalline phase. This lowers the probability for the formation of large polymer crystals and, therefore, depresses the melting point. This effect is described by Equation (8.1).

$$\frac{1}{T_M} - \frac{1}{T_M^0} = -\frac{R}{\Delta H_\mu} \ln N_A , \qquad (8.1)$$

where N_A is the mole fraction of crystallizable groups in a random copolymer.

The behavior of most Saran copolymer systems follows the form of the Flory equation. But VC may be less effective in lowering T_M than are the acrylates. This is illustrated by the plot of $1/T_M$ vs $\ln N_A$ shown in Figure 8.3; of the two comonomer systems shown, the VDC/VC copolymers have the higher melting points at equal comonomer levels.

In copolymers of VDC with long chain alkyl acrylates, the melting point is determined primarily by the mole fraction of

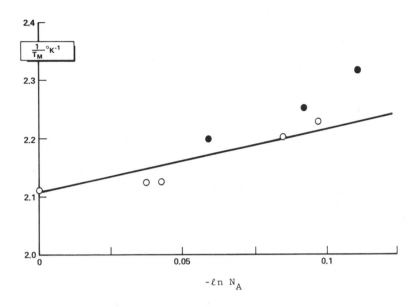

FIGURE 8.3 Effect of comonomer level on melting point of Saran copolymers:
○ VDC/VC copolymers (Ref. 4); ● VDC/MA copolymers (Ref. 3).

acrylate. But if the alkyl group is linear and exceeds 16 carbon
atoms in length, it can crystallize into a separate phase.[32] Then
two melting points are observed, associated with main chain and
side chain crystallization. This is true for both alkyl acrylates and
N-alkyl acrylamides.[33] Copolymers with oleyl side chains do not
show this behavior because the double bond prevents side chain
crystallization.

The relationship between glass transition temperature and copolymer composition is less well understood. For many years,
the Gordon-Taylor theory[34] and its variations were thought to
adequately represent the behavior of most copolymer systems.
They derived the following equation relating T_G to W_A, the
weight fraction of component A in the copolymer:

$$[T_G - T_{GA}] W_A + K[T_G - T_{GB}] W_B = 0 , \qquad (8.2)$$

where T_{GA} and T_{GB} are the glass transition temperatures of the respective homopolymers; $T_{GA} < T_{GB}$ by definition in Equation (8.2). Theoretically, the constant, K, is the ratio of the differences in glass and liquid expansion coefficients:

$$K = \frac{\Delta \alpha_B}{\Delta \alpha_A} = \frac{\alpha L_B - \alpha G_B}{\alpha L_A - \alpha G_A} , \qquad (8.3)$$

where αL and αG are coefficients of thermal expansion above and below T_G. In practice, K has become an adjustable curve fitting parameter with no correlation to the above definition. In many cases, K is near unity and Equation (8.2) reduces to a simple linear expression for T_G as a function of W_A.

$$T_G = T_B + (T_A - T_B) W_A . \qquad (8.4)$$

This is by definition an "ideal" copolymer system. A number of other theories have been proposed, but they lead to essentially the same result.[35,36]

Some Saran copolymer systems fit these simple theories. In most cases, however, the T_G of copolymers is higher than would be predicted and in many cases, passes through a maximum with change in composition.[37] A number of theories have been proposed to account for this phenomenon.[30,38]

The change in T_G with copolymer composition for several common copolymers is shown in Figure 8.4. In every case, T_G increases with comonomer content at low comonomer levels even in cases where the T_G of the other homopolymer is lower. In the latter cases, a maximum T_G is observed at intermediate compositions. In other cases, where the T_G of the homopolymer is much higher than the T_G of PVDC, the glass temperatures of the copolymers increase over the entire composition range.

The change in T_G with composition for copolymers with VC, AN and MMA can be fitted by a Gordon-Taylor type equation.[37] The acrylate and vinyl ester systems, however, require one of the more sophisticated treatments. All of the theories that are capable of predicting maxima fit the data about equally well.[30,37,38]

Illers[39,40] has used Kanig's theory to analyze the VDC/MA

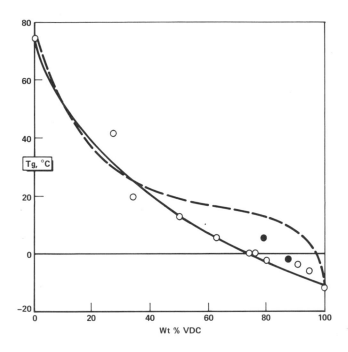

FIGURE 8.4a T_G-composition curve for VDC/VC copolymers (Ref. 37.)
○ Narrow composition distribution specimens; ● broad composition distribution specimens; — curve calculated from Gordon-Taylor equation (with $K =$
0.39); - - - typical curve representing older data.

system. The T_G maximum occurs at 52 mole % MA. He does not
attempt to give a detailed explanation of this result but notes that
the polar interaction between VDC and MA units takes place in
preference to like pair interaction. This condition increases chain
stiffness and gives rise to the the T_G maximum. Both Woodford,[41]
Powell and Elgood[42] and Barton[38] also ascribe the phenomenon to
an increase in chain stiffness. The first author has argued from an
examination of models that the acrylate group interferes sterically

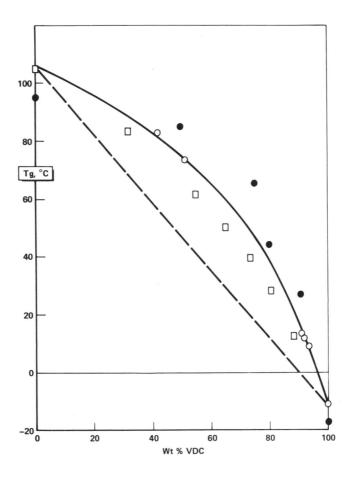

FIGURE 8.4b T_G-composition curve for VDC/AN copolymers (Ref. 37). —— curve calculated from Gordon-Taylor equation with k = 3.0; - - - ideal copolymer.

with free rotation.

Since PVDC acts as a Lewis acid species,[43] we can expect correlations between T_G and comonomer basicity. The category of basic comonomers includes acrylates, methacrylates, AN, etc.

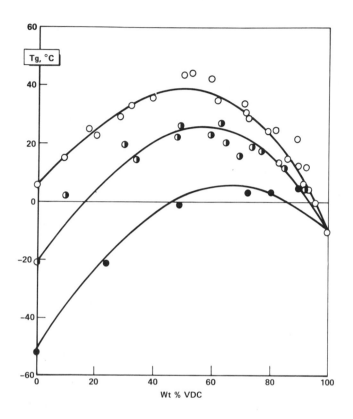

FIGURE 8.4c Effect of side chain length on the T_G of VDC-alkyl acrylate
copolymers (Ref. 37). ○ VDC/MA; ◑ VDC/EA; ● VDC/BA.

Polarity and loss of symmetry in this case both work in the same
direction, to increase T_G.

Polar interactions can take place in two ways: between chains
and between adjacent units on the same chain. It has been
suggested that the latter is responsible for the stiffening effect of
this class of comonomers on the PVDC chain. A dipolar ring
structure can form through a VDC methylene group and the
acrylate carbonyl, as illustrated schematically on the following page:

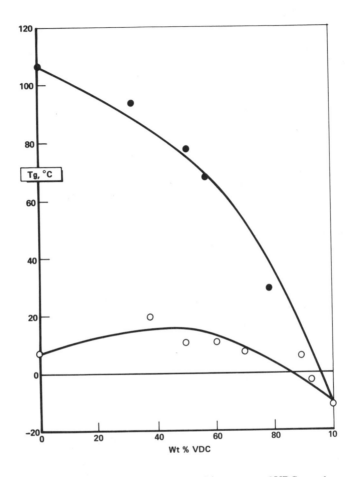

FIGURE 8.4d Comparison of T_G-composition curves of VDC copolymers (Ref. 37). ● VDC/MMA; ○ VDC-VPR data.

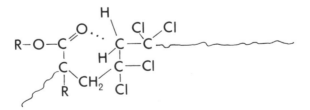

This conformation maximizes contact between the acidic hydrogens and the basic carbonyl oxygen, thus markedly stiffening the polymer chain in the vicinity of the acrylate unit. As noted before, an increase in chain stiffness leads to an increase in T_G.

The maximum in the T_G vs composition plot can be viewed as the result of competing effects. The introduction of acrylate units into a PVDC chain increases stiffness and destroys symmetry. Therefore, T_G rises with comonomer content. But eventually the symmetry of the polymer chain is destroyed completely and all possible polar interactions are satisfied. Further increases in acrylate content will reduce the possibilities for dipole interactions and T_G will fall toward the value of the acrylate homopolymer.

The T_G maximum occurs at approximately 50% in methyl acrylate copolymers but as the length of the alkyl side chain increases, the temperature of the maximum falls and the composition shifts toward higher VDC content. This is a result of side group plasticization being superimposed on the effects described above. The increase in T_G should correlate with the mole fraction of carbonyl functionality. The plasticization effect, however, is dependent on the volume fraction of alkyl substituents in the copolymers.

The limited experimental data available on long chain methacrylate and acrylate copolymers are consistent with this picture.[37,44,45]

A change in composition should also affect T_α. This has been shown in the study by Schmieder and Wolf[9] of the dynamic mechanical properties of VDC/VC copolymers. Their results are summarized in Figure 8.5. Low levels of VC in the copolymer cause both T_α and T_M to drop while T_β is increased, so the net effect of copolymerization in this composition range is to make the polymer softer.

Higher levels of VC, however, increase the T_β above room temperature and the polymer becomes increasingly rigid. In this range, the polymer is essentially amorphous. T_β in amorphous polymers correlates with the heat distortion temperature. This is in contrast to the situation in the low VC range where the heat distortion temperature is a consequence of a high level of crystallinity. Since all copolymers are less crystalline, they tend to soften at lower temperatures (this is probably associated with a drop in T_α

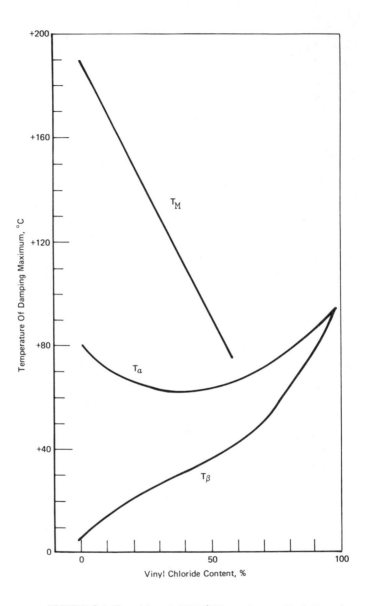

FIGURE 8.5 Transitions in VDC/VC copolymers (Ref. 9).

too). If enough comonomer is introduced, however, the polymer becomes completely amorphous. In the amorphous range, the effect of composition on T_D depends on how T_G varies with composition.

In cases like VDC/VC copolymers where T_G increases with comonomer content, a minimum will be observed in T_D at the composition where the polymer becomes amorphous. This behavior has been demonstrated in the case of VDC/MMA copolymer and VDC/AN copolymers. Thermomechanical softening points for these systems pass through minima. For the MMA copolymers, it falls at 31 °C (68.1% VDC); for the VCN copolymers, at 46 °C (79.4%).[16]

PVDC and its copolymers are non-linear viscoelastic materials. Their mechanical properties have not been well characterized, but as is the case with other crystalline polymers, the observed mechanical response is very dependent on sample history. This makes it difficult to assign a specific value to any given property.

Most of the published data on mechanical properties relate to Saran copolymers of commercial interest. Relatively little has been done with PVDC itself because of the difficulty in fabricating test specimens. The most extensive study is that of Schmieder and Wolf[9] on its dynamic mechanical properties. They measured modulus and loss factor with a torsion pendulum in the range of −150 °C to +190 °C. Their results were shown earlier in Figure 8.1.

The major changes in modulus take place in the glass transition region and as the melting point of the polymer is approached. Below T_G, the modulus is ~ 10^{11} dyn/cm^2. In this temperature range, the polymer is a rigid, brittle solid with low elongation, < 5%. When stressed, it fractures without yielding. From T_G to ~ 130 °C, the modulus decreases slowly with temperature. In this range, the material is ductile with elongation increasing with temperature. The polymer tends to neck when stretched if the molecular weight is high (~ 10^5). Low molecular weight polymers are still brittle in this range.

Above 130 °C, the modulus begins to drop off rapidly; the polymer becomes soft and extensible and can be stretched to 100% elongation or more without necking. The heat distortion temperature as defined by this experiment would fall around 150 °C.

Even at this temperature, the polymer remains a solid material due to the crystallinity.

But when melting begins, at ~185 °C, the polymer changes abruptly from a non-flowing solid to a relatively low viscosity melt[17] and the modulus drops over a narrow temperature range to a very low value. Above 200 °C, the polymer is sufficiently fluid to be extruded or molded.

The mechanical response of PVDC is dependent on other factors besides temperature and molecular weight; of equal importance are degree of crystallinity, orientation and plasticizer level. The high modulus of PVDC at room temperature is due to its high crystallinity; the amorphous polymer is soft and rubbery.[20] Both copolymerization and plasticization reduce crystallinity and therefore act to "soften" the polymer.[46]

PVDC can only tolerate small amounts of plasticizer, of the order of 5% or less. Larger quantities can be worked into the melt bu bleed out when the sample is crystallized. Therefore, plasticization is less effective than copolymerization in reducing crystallinity. The combined approach of copolymerization to reduce crystallinity and plasticization to further soften the polymer is the most effective approach. Copolymers can tolerate much higher levels of plasticizer when crystallized due to their inherently lower crystallinity.

The level of crystallinity in PVDC can also be varied by changing crystallization conditions.[47] Copolymers containing only a small amount of comonomer are similar in this respect to PVDC.[20] The modulus and tensile strength increase with % crystallinity while percent elongation drops.

Tensile properties are also sensitive to polymer morphology. The effect of crystalline texture on the stress strain curves is shown in Figure 8.6. The specimen crystallized at low temperatures with high concentration of nuclei is transparent because of the small crystallite size. It is very ductile and yields and cold draws before failing.

A sample crystallized at high temperature is turbid due to the large size of the spherulites developed under these conditions. It has the same level of crystallinity but a coarser texture. When stretched it yields but does not draw and hence has lower elongation.

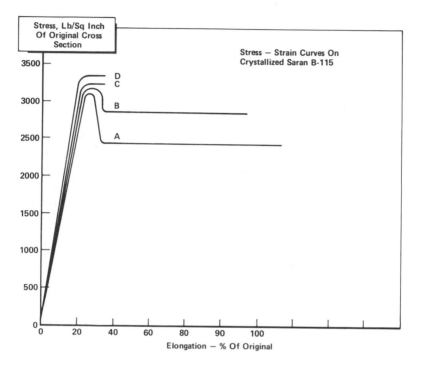

FIGURE 8.6 Effect of thermal history on the stress-strain curves for a VDC/VC copolymer. All samples melted at 200 °C, quenched to 10 °C, aged and then crystallized at 140 °C. A – aged 14 hours at 10 °C; B – aged 18 seconds at 50 °C; C – aged 6 seconds at 90 °C; D – aged 35 seconds at 130 °C.

(Courtesy R. M. Wiley, The Dow Chemical Company.)

Orientation dramatically improves the mechanical properties of Saran.[17,45] This is also due to morphological differences. Either polymer chains or crystalline regions (or both) can be aligned. When tested parallel to the direction of orientation, such specimens are much stronger. The effect of uniaxial orientation is shown by the data in Table 8.4. In these experiments, an increase in draw ratio indicates an increase in orientation.[48]

TABLE 8.4

Effect of Stretch Ratio Upon Tensile Strength and % Elongation
(Average of 5 determination, using the Instron test at 2 in./min)

Stretch ratio	Tensile strength, psi	% Elongation
2.50 : 1	34,080	23.2
2.75 : 1	33,960	21.7
3.00 : 1	43,960	26.3
3.25 : 1	38,900	33.1
3.50 : 1	45,820	19.2
3.75 : 1	47,830	21.8
4.00 : 1	46,380	19.7
4.19 : 1	45,520	16.2

Uniaxial orientation increases strength in the parallel direction but usually reduces it in the perpendicular direction. Specimens can also be biaxially oriented so that the strength is improved in both directions. Some properties of a typical biaxially oriented film of a plasticized Saran B resin are shown in Table 8.5.[49] As in the case of monofilaments, the increase in tensile strength is accompanied by a drop in elongation.

TABLE 8.5

Properties of Biaxially Oriented Saran B

Property	Machine direction	Transverse
Tensile strength, lb/in.^{2}	10,600	16,000
% Elongation	55	35
Yield strength, lb/in.^{2}	10,000	None
Modulus. lb/in.^{2}	7.0×10^4	5.0×10^4

The Saran B resins used commercially are nearly always plasticized. Havens[50] has made a broad study of the effect of type and amount of plasticizer on both mechanical and rheological properties. He found that the stiffness of the polymer varied with plasticizer concentration at 40 °C according to Equation (8.5)

$$\log M_f = \log M_0 - K_m B , \tag{8.5}$$

where M_f is Young's modulus of the unplasticized polymer, M_0 is Young's modulus without plasticizer, B is % plasticizer and K_m is a constant characteristic of plasticizer effectiveness.

He also observed that the flex temperature (which is related to heat distortion temperature) changed linearly with plasticizer level.

Plasticization shifts the modulus-temperature curves to a lower range.[51] When the data are plotted on a semilog scale, as shown in Figure 8.7, a sharp break in the curve becomes evident. This corresponds to the temperature at the upper end of the glass transition region and, therefore, falls approximately 10 °C above T_G. At lower temperatures, the modulus levels off to a value characteristic of the glassy state.

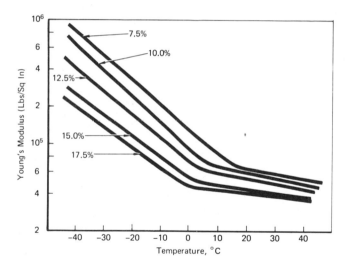

FIGURE 8.7 Effect of varying plasticizer B content in Saran B (Ref. 51).

Plasticizer concentration has the greatest effect on the modulus in the transition region. It reduces both T_G and the brittle temperature. At temperatures well above or below T_G, the effect is less pronounced though in all cases the modulus decreases with increasing plasticizer level.

The overall shape of the modulus temperature curve is not greatly affected by plasticizer level. But it is very sensitive to type of plasticizer in the transition region. A good plasticizer, like α-chloronaphthalene, causes a more rapid drop in modulus at T_G to a lower value than that caused by a poor plasticizer, like butyl stearate.

The mechanical properties of Saran copolymers are dependent on the type and amount of comonomer. The more common copolymers containing around 10% acrylonitrile or lower alkyl acrylates tend to be of lower crystallinity than Saran B resins. But because of their higher glass transition temperatures, they are not significantly softer.

TABLE 8.6

Mechanical and Thermal Properties of Copolymers of Vinylidene Chloride and Alkyl Acrylates (Ref. 45)

Acrylate content Copolymer		Tensile strength	Elonga-tion	100% Modulus	T_G °C
Mole %	Wt. %	lb/in.^{2}	%	lb/in.^{2}	
		Butyl acrylate polymers			
7.5	11.0	2600	45	–	12.5
10.1	14.3	2560	90	–	8.5
12.5	18.3	1930	200	1860	7.0
15.5	22.7	930	400	600	2.0
22.0	32.2	590	500	340	−1.5
		Octyl acrylate copolymers			
5.0	10.2	3750	140	–	–
7.6	15.5	2690	280	2660	0.0
10.1	20.7	1820	430	1740	−3.5
12.9	26.4	1350	500	1180	−8
15.2	32.3	715	500	550	−11.5
20.8	42.5	170	500	110	−16
		Octadecyl acrylate copolymers			
5.3	17.7	4060	110	3950	−7
8.5	28.4	2250	425	2080	−11
11.9	39.8	1700	500	1595	−16
13.2	44.1	1320	500	1190	−18
16.0	53.5	900	430	770	−21.5

The higher alkyl acrylate copolymers have substantially different properties. Jordan and coworkers[45] have made an extensive study of these copolymers. Their results are summarized in Table 8.6. The tensile properties correlate with mole % acrylate showing that the dominant factor in the composition range investigated is crystallinity.

Based on weight fraction, fairly high levels of the long chain comonomer, n-octadecyl acrylate can be incorporated without destroying crystallinity. Therefore, the tensile strength is significantly higher. The crystalline phase is nonetheless being diluted out by the waxlike alkyl side chains so the modulus still drops.

Jordan and coworkers also evaluated the N-alkylacrylamide copolymers.[33] The effect of this type of comonomer on mechanical properties of Saran is similar to that described for acrylates. But the amide functionality seems to accentuate brittleness probably due to stronger polar interaction along the chain backbone.

Brittle failure in Saran is more sensitive to factors like molecular weight and plasticizer level than it is to type of comonomer. Boyer and Spencer[8] found that the brittle points of PVDC followed the equation:

$$\frac{1}{T_B} = A + BM^{1/2} . \tag{8.6}$$

Many other polymers show a similar relation between brittle point and molecular weight.

In crystalline polymers, the brittle point is also sensitive to morphology and degree of crystallinity. Highly crystalline, carefully annealed specimens of even a tough polymer like polyethylene are very brittle. The same is true of Saran. The toughest materials are obtained by crystallizing under quenching conditions.[52] Even in this case, however, impact strength falls with increasing crystallinity and the polymer can undergo brittle failure at temperatures well above T_G. The amorphous polymer does not fracture under the same conditions.

Saran resins are commonly plasticized to reduce brittleness at low temperatures. The effect of a good plasticizer is to shift the brittle point to a lower temperature thereby increasing the range over which the polymer has good impact strength.[50]

The effect of comonomer level on impact strength and brittle point depends on its structure as noted earlier. Saran B resins have better room temperature impact strength than other crystalline copolymers because of their lower T_G. The brittle points of the higher n-alkyl acrylate copolymers are lower, however, which suggests that they might be tougher materials.[45]

The factors that determine impact strength in crystalline polymers are not as well understood as in the case of glassy polymers. But Wada and Kasahara[53] have shown that in general the impact properties of many polymers can be correlated with dynamic mechanical properties. They plotted impact vs area under the loss peak for a variety of materials and found that the points seem to fall into two categories: The tough polymers were those with T_G below the test temperature. Brittle polymers were those with high T_G. In spite of its low T_G, PVDC falls with the latter group suggesting that its brittle point may be very sensitive to test rate. This was already noted earlier in comparing T_G by various test methods. The T_G of PVDC at the frequency of an impact test (approximately 100 hertz) is well above room temperature. Copolymers respond similarly.

REFERENCES

1. T. Alfrey, Jr. and E. F. Gurnee, *Organic Polymers*, Prentice Hall, N.Y. (1969).
2. R. A. Wessling, J. H. Oswald and I. R. Harrison, *J. Polym. Sci. Phys.*, **11**, 875 (1973).
3. K. Okuda, *J. Polym. Sci. A*, **2**, 1749 (1964).
4. D. Kockott, *Kolloid-Z. Z. Polym.*, **206**, 122 (1965).
5. B. V. Lebedev, I. B. Rabinovich and V. A. Budarina, *Polym. Sci. USSR*, **9**, 545 (1967).
6. R. F. Boyer and R. S. Spencer, *J. Appl. Phys.*, **15**, 398 (1944).
7. R. F. Boyer and R. S. Spencer, *J. Appl. Phys.*, **16**, 594 (1945).
8. R. F. Boyer and R. S. Spencer, in H. Mark and G. S. Whitby (eds.), *Advances in Colloid Science*, Vol. II, Interscience, N.Y., pp. 1-55 (1946).
9. K. Schmieder and K. Wolf, *Kolloid-Z. Z. Polym.*, **134**, 149 (1953).
10. R. W. Eykamp, A. M. Schneider and E. W. Merrill, *J. Polym. Sci. A-2*, **4**, 1025 (1966).

11.   R. A. Wessling and F. G. Edwards, *Enc. Polym. Sci. Tech.*, **14**, 540 (1971).
12.   R. Simha and R. F. Boyer, *J. Chem. Phys.*, **37**, 1003 (1962).
13.   The Dow Chemical Company, unpublished data.
14.   N. G. McCrum, B. E. Read and G. Williams, *Anelastic and Dielectric Effects in Polymer Solids*, Wiley, N.Y., p. 436 (1969).
15.   T. Bunkyo-ku, *Rep. Prog. Polym. Phys., Japan*, **7**, 271 (1964).
16.   G. S. Kolesnikov, L. S. Fedorova, B. L. Tsetlin and N. V. Klimentova, *Bull. Acad. Sci. USSR, Dev. Chem. Sci.*, **1959**, 701 (1959).
17.   R. C. Reinhardt, *Ind. Eng. Chem.*, **35**, 422 (1943).
18.   R. A. Wessling and I. R. Harrison, *J. Polym. Sci. A-1*, **9**, 3471 (1971).
19.   J. D. Hoffman, *SPE Trans.*, **4**, 315 (1964).
20.   R. D. Lowry and R. M. Wiley, The Dow Chemical Company, unpublished results.
21.   M. L. Miller, *The Structure of Polymers*, Reinhold, N.Y., p. 291 (1966).
22.   S. Saito and T. Nakajima, *J. Polym. Sci.*, **37**, 229 (1959).
23.   J. Heller and D. J. Lyman, *Polym. Lett.*, **1**, 317 (1963).
24.   Y. Ishida, M. Yamamoto and M. Takayanagi, *Kolloid-Z. Z. Polym.*, **168**, 124 (1960).
25.   R. F. Boyer, *Rubber Rev.*, **36**, 1303 (1963).
26.   P. J. Flory, *Principles of Polymer Chemistry*, Cornell University Press, Ithaca, N.Y., p. 53 (1953).
27.   W. A. Lee and G. J. McKnight, *Brit. Polym. J.*, **2**, 73 (1970).
28.   F. Wurstlin, *Kolloid-Z. Z. Polym.*, **120**, 84 (1951).
29.   J. H. Gibbs and E. A. DiMarzio, *J. Chem. Phys.*, **28**, 373 (1958); 807 (1958).
30.   G. Kanig, *Kolloid-Z. Z. Polym.*, **190**, 1 (1963).
31.   P. J. Flory, *Principles of Polymer Chemistry*, Cornell University Press, Ithaca, N.Y., Chapter 13 (1953).
32.   T. Alfrey, Jr., N. Wiederham, R. S. Stein and A. Tobolsky, *J. Colloid Sci.*, **4**, 211 (1949).
33.   E. F. Jordan, G. R. Riser, B. Artymyshyn, W. E. Parker, W. J. Pensabene and A. N. Wrigley, *J. Appl. Polym. Sci.*, **13**, 1777 (1969).
34.   M. Gordon and J. S. Taylor, *J. Appl. Chem.*, **2**, 492 (1952).
35.   E. A. DiMarzio and J. H. Gibbs, *J. Polym. Sci.*, **15**, 121 (1959).
36.   R. A. Hayes, *J. Appl. Polym. Sci.*, **5**, 318 (1961).
37.   R. A. Wessling, F. L. Dicken, S. R. Kurowsky and D. S. Gibbs, *Appl. Polym. Symp.*, **25**, 83 (1974).
38.   J. M. Barton, *J. Polym. Sci. C*, **30**, 573 (1970).
39.   K-H. Illers, *Kolloid-Z. Z. Polym.*, **190**, 16 (1963).
40.   K-H. Illers, *Ber. Bunsenges. Phys. Chem.*, **70**, 353 (1966).
41.   D. M. Woodford, *Chem. Ind.*, **1966**, 316 (1966).
42.   E. Powell and B. G. Elgood, *Chem. Ind.*, **1966**, 901 (1966).
43.   R. A. Wessling, *J. Appl. Polym. Sci.*, **14**, 1531 (1970).
44.   The Dow Chemical Company, unpublished data.

45. E. F. Jordan, W. E. Palm, L. D. Witnauer and W. S. Bort, *Ind. Eng. Chem.*, **49**, 1695 (1957).
46. R. M. Wiley, U.S. 2,160,932, to The Dow Chemical Company (1939).
47. R. M. Wiley, U.S. 2,329,571, to The Dow Chemical Company (1942).
48. E. D. Serdynsky, in H. Mark, S. M. Atlas and E. Cernia (eds.), *Man-Made Fibers*, Vol. 3, Interscience, N.Y., p. 303 (1968).
49. J. Jack, *Brit. Plastics*, **34**, 391 (1961).
50. C. B. Havens, *Ind. Eng. Chem.*, **42**, 318 (1950).
51. H. W. Moll and W. J. LeFevre, *Ind. Eng. Chem.*, **40**, 2172 (1948).
52. R. M. Wiley, U.S. 2,205,449, to The Dow Chemical Company (1938).
53. Y. Wada and T. Kasahara, *J. Appl. Polym. Sci.*, **11**, 1661 (1967).

CHAPTER 9

Degradation

PVDC is highly resistant to oxidation and biodegradation which makes it extremely durable under ambient conditions. But it decomposes readily when heated. The thermal decomposition reaction has been widely studied. But this is not the only mode of degradation to which PVDC is susceptible; radiation, acids, and strong bases can also be used to induce decomposition even at ambient temperatures. A common feature of all degradation processes is the elimination of chlorine from the polymer as either chloride ion or HCl gas. Monitoring the release of either product is a convenient method of following the kinetics of these reactions.

9.1   THERMAL DEGRADATION

9.1.1   Degradation in the solid state[1-10]

PVDC begins to evolve HCl at ~120 °C. The rate of evolution increases rapidly with temperature. The only volatile product detectable below 190 °C is HCl. Decomposition over this range takes place in the solid state.

The polymer contains ~ 20.6 milliequivalents of HCl/gram of polymer. But only about half of the available acid is released in the solid state reaction. A typical plot of HCl evolved as a function of time at constant temperature is shown in Figure 9.1.[10] The rate increases with time initially, reaching a maximum at ~10% reaction. (Based on % of total HCl evolved.) From 10 to 40% reaction, the data fit a first order plot as shown in Figure 9.2. The rate falls rapidly to zero at higher extents of reaction. The limiting yield of

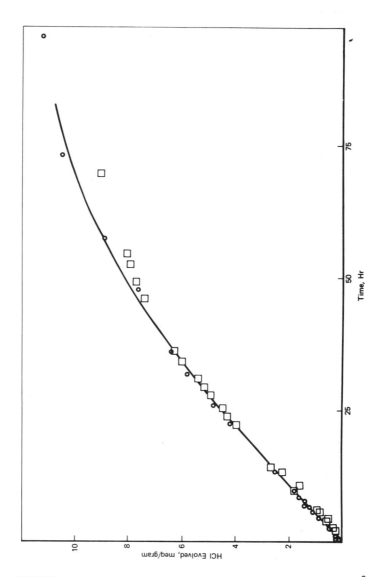

FIGURE 9.1 Thermal decomposition of mass polymerized PVDC at 160 °C
under vacuum (Ref. 10).

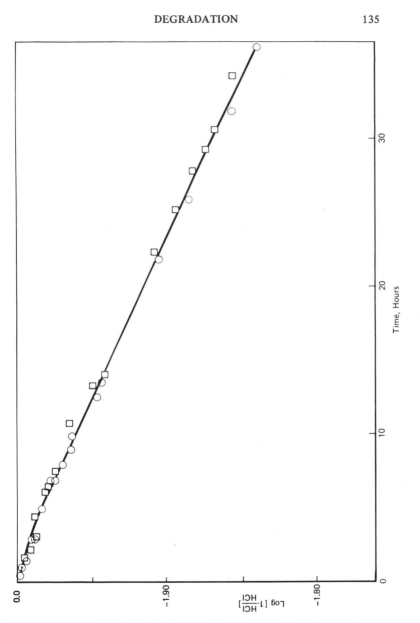

FIGURE 9.2 A first order plot for the thermal decomposition of mass polymerized PVDC (Ref. 10).

HCl appears to increase with temperature but does not exceed ~ 60% at temperatures below 190 °C.

Degradation studies are commonly carried out on "as polymerized" powders. The powder morphology seems to be retained in the degraded products when the reaction is carried out in the solid state.[11,12] But in the same temperature range where solid state decomposition occurs, the polymer is also annealing and recrystallizing to a state of higher crystallinity.[13] These processes are competitive initially but the crosslinking of the polymer which accompanies thermal degradation quickly freezes in the gross morphology before changes are apparent.

Crosslinking and discoloration can be detected at a very early stage of decomposition. Substantial crosslinking is observed at < 1% HCl evolved.[7] The polymer becomes completely insoluble at ~ 10% reaction, and after 30% of HCl is evolved, the last traces of PVDC crystallinity disappear and the degradation product becomes infusible. These changes are reflected by the decrease in melting point with extent of reaction as shown in Table 9.1.[13] In the absence of degradation, annealing would be expected to cause the melting point to increase.[14]

The non-volatile degradation products have been characterized mainly by IR spectroscopy.[7,15] Isolated double bonds are first observed, but by 5% reaction, conjugated sequences can be detected. The polymer gradually discolors changing from white to yellow to brown. This has been attributed to the formation of long polyene sequences.[16] A set of IR spectra taken at different stages of decomposition (Figure 9.3) show these changes quite clearly. While there is general agreement that the degradation product at ~ 50% HCl loss is crosslinked polychloroacetylene, there are differing views on the effect of molecular weight and the order of the reaction. Burnett and coworkers[7] found a simple inverse dependence of rate of decomposition on molecular weight. This suggested that the reaction was initiated at chain ends. But they analyzed only the early part of the reaction and assumed zero order kinetics.

A number of other workers have also interpreted the reaction as a zero order process,[4,17] even though the rate is not constant.

PVDC is converted to polychloroacetylene after only 50%

TABLE 9.1
Effect of Annealing at 160 $^\circ$C on Melting Point of PVDC (Ref. 13)

Exposure time, h	% Decomposition[a]	T_M, $^\circ$C
0	0.0	200
1.25	0.92	197
2.25	1.04	196
4	2.72	192
12	11.9	167
24	21.4	b
48	37.4	—

[a] (gHCl released)/(gHCl available) $\times$ 100.
[b] No observable melting point.

reaction (based on HCl loss). The data follow a first order plot over more of this range better than zero order. The activation energies obtained from an Arrhenius plot of first order rate constants are around 34 kcal/mole.[8,10]

There is no significant effect of either molecular weight or end group structure on the value of the first order rate constant; this suggests that molecular weight is involved only in determining the initial rate. And even here, it may be true only for chains with certain types of end groups.

The role of morphology in the reaction was not considered until Bailey and Everett[11,12] noted that carbon powders derived from Saran were similar in appearance to the starting polymer powder. Small particle size powders degrade more rapidly.[8] This appears to be associated with more subtle differences than surface area. The effect of morphology can be obscured by other factors, also. For example, polymers prepared in aqueous media appear to have slightly higher rates of decomposition than mass polymerized PVDC.[10] And while there are very large differences in morphology between these samples, differences in stability are more likely to be associated with slight chemical changes introduced into the polymer due to contact with the polymerization media.

The rate of decomposition of lamellar PVDC crystals was studied in order to get a better picture of the effect of morphology,[10] but recrystallization from polar solvents sensitized the

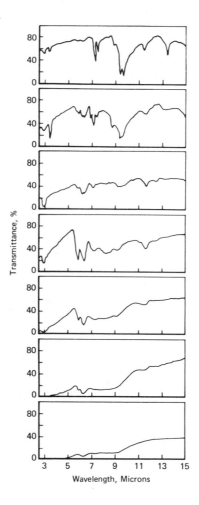

FIGURE 9.3 Infrared spectra using KBr pellet technique. Top to bottom: pure Saran, 20.0%, 42.0%, 56.0%, 66.4%, 70.0% and 75.3% weight loss respectively (Ref. 15).

polymer to subsequent thermal decomposition. This had a levelling effect on the results; all recrystallized polymers degraded much faster but differences between samples were not clear cut. Though annealing would be expected in the temperature range studied, this effect could not be isolated.

Most of the above studies were made either under an inert gas or *in vacuo*. There is little difference between these environments. But it has been noted in several instances that oxygen does affect the course of the reaction.

These results suggest that decomposition occurs by a radical reaction.

The likelihood that a free radical mechanism was involved was postulated much earlier.[4] This hypothesis was further confirmed by studies with inhibitors[7] and by ESR spectroscopy.[18]

A qualitative mechanism for the thermal dehydrochlorination of PVDC was proposed many years ago by workers at the Dow Chemical Company, particularly Boyer.[1,2,16] They visualized a two stage reaction as described below:

1) Unzipping

$$\text{+}(CH_2-CCl_2\text{+})_n \xrightarrow{\Delta} \text{+}(CH=CCl\text{+})_n + nHCl$$

2) Crosslinking

$$\text{+}(CH=CCl\text{+})_n \xrightarrow{\Delta} \text{Crosslinked aromatic products.}$$

The mechanism for the first stage involved an initial formation of a double bond. The first double bond initiated an autocatalytic sequential HCl elimination to form a conjugated polyene. The driving force for a chain reaction was ascribed to allylic activation of an undegraded unit by an adjacent double bond.

The crosslinking reactions and the second stage processes were assumed to involve Diels-Alder condensation to form rings which could undergo further dehydrochlorination. This aspect of the mechanism was described in greater detail by Winslow and co-workers.[3]

Their data also supported a two-step mechanism with only the first stage being important below 200 °C. They pointed out the possibility that initiation was not a radical process even though the "unzipping" reaction apparently was.

A continuing study of the solid state thermal degradation by many different groups has confirmed these early conclusions and has added more specific details to the mechanism. But the observations on the effect of morphology and polar solvents on the kinetics have yet to be incorporated in a satisfactory manner.[10]

The evidence suggests that an active site forms on the surface of a polymer crystal. A free radical chain reaction is then initiated at this site and propagates through the crystal. The length of the conjugated sequence and the yield of HCl per initiation must therefore be related to the crystal dimensions.

Reaction in the crystalline phase may require temperatures of at least 120 °C. This is in the range where PVDC crystals begin to soften and anneal.[13] The thickness of the lamellar crystals (which are probably the basic component of morphology for all of the "as polymerized" powders) should increase with annealing time to a characteristic value which is proportional to temperature. But as mentioned earlier, this process is blocked early in the reaction by crosslinking.

The interaction between annealing and crosslinking probably determines the nature of the induction period. This may be why molecular weight affects the early part of the reaction. The higher the molecular weight, the faster the polymer crosslinks.

Once crosslinking reaches a certain point, annealing stops; or possibly at low temperatures, the maximum crystal thickness is reached first.

The reaction then apparently passes into the first-order region. This mechanism would predict that crosslinking at room temperature followed by heating should shorten the induction period and reduce the first-order rate. This is the result observed by Everett and Taylor.[9]

The nature of the initiation reaction is still a matter of speculation. Everett *et al.*[8,9] favor thermal dissociation of a C—Cl bond. The resulting radical can either couple with a radical on an adjacent chain to form a surface crosslink or propagate into the crystal to form a polyene sequence. Local stereochemical differences in the chain folds were assumed to account for the relative proportions of crosslinking and propagation.

Another explanation is that the active sites are double bonds that form on the crystal surface. Double bonds can be formed by a number of polar mechanisms such as solvent induced E2 elimination[10,19] or a four center unimolecular polar elimination.[20] They can also be photochemically generated.[9]

9.1.2 Degradation of Saran copolymers and related halogenated polymers

It was quickly recognized by the early Dow workers that some of the copolymers they prepared were more stable than PVDC. Boyer[16] proposed that the greater stability of acrylate copolymers was related to the fact that long VDC sequences were broken up by comonomer units. This limited the potential length of an "unzipping" reaction. He was able to get a reasonable correlation between VDC sequence length and light stability in support of this hypothesis.

The greater stability of copolymers of VDC with VC over PVDC cannot be attributed to that mechanism since both monomer units dehydrochlorinate. Differences here may be due instead to the greater stability of the C–Cl bond on a VC unit since it is known that secondary halides are less reactive than tertiary halides.

In agreement with this assumption, the rates of dehydrochlorination of these copolymers fall progressively with increasing VC content.[21] These composition changes are accompanied by a major change in morphology. With increasing VC content, the morphology changes from highly crystalline lamellar particles to nearly amorphous spherical particles. From the speculation about the effect of morphology on the kinetics discussed earlier, we would expect the latter to be more stable regardless of composition differences.

If the ease of degradation of PVDC is associated with the stability of the C–Cl bond, then comonomers with polar functional groups such as $-C{\equiv}N$ should reduce polymer stability. But even more specific neighboring group effects might be anticipated. The cyclization reaction in VDC/MMA copolymers described by Zutty and Welch[22] is a case in point. The ester group reacts at high temperatures with an adjacent VDC unit to form a lactone.

Burnett *et al.*[23] studied the kinetics of decomposition of several copolymers of VDC including the above system. Their rate data show that the VDC/MMA copolymers degrade faster than the homopolymer. The HCl evolution was first order in the range 120-200 °C with an activation energy of 27 ± 1.5 kcal/mole.

Methacrylonitrile copolymers evolved only HCl in this temperature range and also followed first-order kinetics, but the activation energy was 37 ± 1.5 kcal/mole. Though rates of decomposition were higher, Burnett *et al.* did not see a neighboring group effect like that in the MMA copolymer.

A different kind of neighboring group effect was observed in VDC/styrene copolymers.[24] The kinetic analysis of the reaction was complicated by what appeared to be retardation of the reaction by some of the products. Again HCl was the volatile product and rates of decomposition were much higher than that of PVDC. The evidence was in favor of a radical mechanism with initiation at VDC units located adjacent to styrene units.

The above studies relate to random copolymers. Dolezel *et al.*[17] have studied the rates of degradation of PVDC blended with and grafted to ethylene/propylene and butadiene rubbers. The presence of the rubbery component lowered the rate of decomposition and increased the activation energy.

Chemical differences must, of course, be an important factor in the stability of halogenated polymers. The type of halogen substituent, for example, is the dominant factor. In line with the stabilities of low molecular weight halocarbons, the order of stabilities of poly vinyl and vinylidene halides is:

PVF > PVC > PVB
PVDF > PVDC > PVDB

A copolymer of a chloride and a bromide is less stable than the pure chloride. Bromine units probably act as initiating sites for decomposition.

Less predictable results are encountered when the arrangement of the halogen substituents is altered. Head to head PVDC for example is more stable than the normal product. Murayama and Amagi[25] prepared this polymer by chlorination of polydichlorobutadiene.

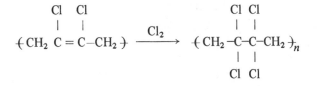

The TGA curves at 2.5 °C/min in argon are shown in Figure 9.4.
Head-to-head PVDC does not start losing weight until ∼ 250 °C.
Rather than being associated with changes in the unzipping type
sequential HCl elimination as proposed, it could also be related to
steric effects. As Frevel[26] suggested many years ago, the severe
crowding of the Cl atoms in the 1,3 disposition along the chain may
sensitize the polymer to dehydrochlorination. The head-to-head
structure is less sterically hindered,[27] and should not be affected.
In both cases, however, the formation of one double bond activates
an adjacent unit to further dehydrochlorination.

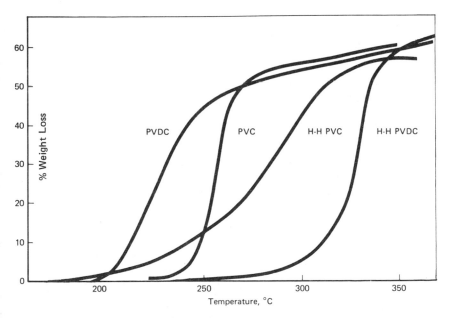

FIGURE 9.4 TGA curves showing the difference in stability between head-to-
head PVDC and normal PVDC (Ref. 25).

9.1.3   Degradation in the molten state

The above discussions relate to degradation in the solid state under isothermal conditions. In many cases, however, degradation is carried out by heating the specimen at a constant rate. The polymer usually melts without degradation using the normal heating rates of 5-30 °C/min in a DTA experiment. Because of the induction period in the dehydrochlorination reaction, the polymer can be heated rapidly up to ~ 225 °C before serious decomposition starts (see, for example, Figure 8.2).

A combination DTA-TGA gas evolution experiment[28] shows the absence of degradation below T_M. Repeated cycling through the melting point and recrystallizing results in a gradual drop in T_M and discoloration of the polymer. This means that the polymer can not be melted and recrystallized if a completely undegraded specimen is desired.

The decomposition above T_M takes off almost explosively. The polymer foams and turns black as it releases a large quantity of HCl gas. The foam undergoes crosslinking to a rigid char.

Only about 60% HCl is evolved in the 200-225 °C range. Heating to higher temperatures is still necessary to get complete dehydrochlorination. The reactions that take place above the melting point are not as well defined as the solid state reactions.

Still another sequence of reactions can take place if a sample of PVDC is rapidly injected into a furnace at a temperature around its ceiling temperature, 350 °C.[29] In this temperature range, the rate of depolymerization is competitive with the rate of dehydrochlorination. Therefore, a variety of new volatile products are obtained. The major ones are vinylidene chloride and 1,3,5 trichlorobenzene. Higher chlorinated condensed ring aromatic structures are also detected. The yield of non-volatile carbonaceous residue is correspondingly reduced.

The kinetics of the process have not been investigated; it has been used primarily as an analytical tool for measuring compositions of VDC/VC copolymers. The distribution of products indicates that the reaction is sensitive to differences in copolymer microstructure. It is an open question yet whether the aromatic products are formed by cyclization of polyene fragments as suggested or by cyclic trimerization of the monomeric products in the vapor phase.

9.2   OTHER MODES OF DEGRADATION

9.2.1   Radiation induced degradation

Pure poly(vinylidene chloride) does not degrade significantly in the dark at room temperature. However, photodegradation can be initiated by ultraviolet light with wavelengths below ~ 4000 Å.[1] Ionizing radiation also degrades PVDC, but the course of the reaction is quite different.[30]

During photodegradation, the polymer gradually crosslinks while the color changes from yellow to brown. PVDC powders treated continuously with UV for several weeks lose about 5% of the available HCl.[9] The irradiated powders behave differently in subsequent thermal degradation. The induction period is shortened and the first order rate constant is reduced by ~ 25%. This is probably associated with the fact that the polymer was crosslinked before heating, thus preventing the annealing process from taking place.

The absorption of light by powders is quite inefficient because of scattering. Therefore, quantitative studies have been carried out on transparent films. The light stability of a number of copolymers was studied at the Dow Chemical Company. Boyer[16] used these results in formulating his ideas about the reaction mechanism. He found that acrylate copolymers had better light stability than vinyl chloride copolymers. The stability increased with acrylate content. Stability in this case was correlated with high transmission of light in the degraded films.

Oster *et al.*[31,32] made a detailed study of the irradiation of a Saran B1500 film. Irradiation in vacuo with 254 mμ light produced a yellow-brown color. The UV and visible spectra of the degraded films were recorded at different exposure times. The curves are shown in Figure 9.5.

The single broad absorption band at 285 mμ was the only feature in the spectrum. This band increased in intensity with exposure time but did not shift in wavelength. This suggests the formation of a single type of absorbing species whose concentration increases with exposure time.

In order to fit Boyer's hypothesis this would have to be a conjugated polyene with a narrow sequence length distribution.

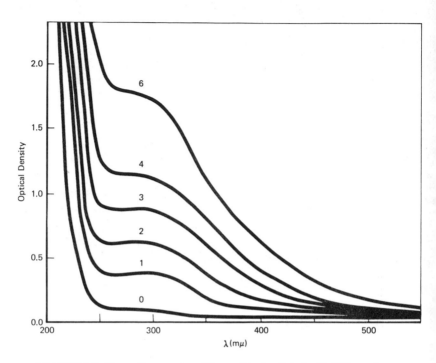

FIGURE 9.5 Absorption spectra of Saran film irradiated with 254 mμ for different times (0-6 h as indicated on graph). Intensity 5×10^{18} quanta/cm^2/h (Ref. 31).

Another possibility suggested by Oster *et al.* was a chlorinated aromatic ring structure. They ruled out the possibility of a conjugated carbonyl because the samples discolored more severely *in vacuo* than in an oxygen environment. In order to get a reasonable estimate for the extinction coefficient for the chromophore, they proposed that it involved interaction between a conjugated system and unpaired electrons. In support of this hypothesis, they observed that irradiated films produced a strong ESR signal.

Oster *et al.* also observed that the film became increasingly insoluble in boiling dioxane with exposure time. Kryszewski and Mucha[33] studied this phenomenon further. The gel fraction increased with exposure time to a maximum of $\sim$ 70%. This was accompanied by a drop in film density and an increase in T_G.

Ionizing radiation can cause either crosslinking or chain scission as well as dehydrochlorination.[30] Everett and Taylor[9] found that irradiation *in vacuo* of PVDC powder with a CO^{60} source at 7.6 rad/h. caused chain scission only. They observed no discoloration or HCl release. But heavily dosed samples showed a high concentration of unpaired electrons and degraded more rapidly when heated.

Harmer and Rabb[34] made a more extensive study of γ-ray treatment. Their results are shown in Table 9.2. PVC crosslinks under the conditions they used, and the addition of VDC units in the chain seems to promote chain scission. The effect observed was dependent on morphology. In all cases, "as polymerized" powders decreased in molecular weight. Crosslinking was detected only in the higher VC polymers in the molded sheet form. Both an increase in specimen temperature during irradiation and the presence of oxygen promoted crosslinking. These results suggest that the crosslinking reaction takes place in amorphous regions by a free radical mechanism.

TABLE 9.2

Effect of Physical Form on Crosslinking of VDC/VC Copolymers (Ref. 34)

Vinylidene chloride in copolymer, %	Physical form	Change in viscosity, %
89	Powder	− 4.6
89	Molding	− 1.8
73	Powder	−13.6
73	Molding	+ 18.4
15	Powder	− 3.7
15	Molding	+ 9.7

Evacuated, sealed samples, irradiated at ambient temperature for 6.3 Mrads, dose rate of 400 krads/h.

9.2.2 Base induced degradation

PVDC can be degraded by the action of almost any basic reagent. These reactions are commonly heterogeneous with the polymer present as a solid phase. Therefore, the rate is controlled by polymer surface area and the ability of the reagent to penetrate the polymer phase.

Aqueous caustic solutions, for example, have little effect on PVDC at room temperature because water neither wets nor swells the polymer. Therefore, hydroxide ion, the active species in this system, is unable to get at the polymer. But in the presence of a wetting agent, surface degradation can occur.[35] Even then, penetration into the polymer phase is very slow.

The reactivity of a base depends both on its base strength and its compatibility with PVDC. Weak bases, like polar aprotic solvents, swell the polymer readily and in some cases even dissolve it. But they are not basic enough to induce dehydrochlorination except at elevated temperatures.

Amines have the best combination of basicity and solubility. Therefore, they make very effective dehydrochlorinating agents. Butyl amine has been found to be very reactive; it attacks the surface of Saran film vigorously at room temperature.[36] The reaction is apparently confined to the surface because of slow diffusion of the amine through the degraded layer.

The possibility that crystalline and amorphous phases might react differently was not considered in the above study. Harrison and Baer[19] however, showed that segments of the polymer chain within a crystal are not attacked by pyridine in a non-swelling medium.

Dehydrochlorination of lamellar crystals suspended in an alcoholic solution of pyridine takes place rapidly at first but then levels off to a very slow rate. The amount of HCl removed in the fast part of the reaction is equivalent to dehydrochlorination of two units per chain fold.

The reaction appeared to take place in a stepwise fashion. Isolated double bonds were first observed. Later, an increasing concentration of conjugated pairs of double bonds could be detected. Longer sequences appeared only after extended reaction.

The results suggest that pyridine attacks the polymer in a chain fold region inducing an **E2** elimination. Further reaction is apparently restricted by steric factors even though allylic activation should promote sequential elimination. **E2** eliminations take place more readily when the leaving groups are trans to each other. The constraints of the polymer chain and the crystal lattice probably limit the occurence of this conformation.

When a stronger base is used, even in a non-swelling medium, the reaction is not confined to the polymer surface. Barton and coworkers[37] found that solutions of KOH is isopropyl alcohol could strip the polymer free of chlorine in less than 12 h. The kinetic pattern was similar to that described by Harrison and Baer, a fast initial reaction followed by a much slower secondary process. This seems to be indicative of a diffusion-controlled process.

The mechanism by which base induced dehydrochlorination proceeds beyond 50% is not completely defined. Barton *et al.* hypothesized that the initial reaction formed polychloracetylene. This was followed by a slow displacement of Cl by alkoxide which after rearrangement and olefin elimination left a hydroxyl substituted polyene. In support of this mechanism, they reported that the final product contained substantial amounts of bound oxygen.

The possibility that acetylenic bonds were formed and later oxidized fits better with the results of Tsuchida and coworkers.[38,39] They found that $NaNH_2$ in liquid NH_3 was a better base than alkoxide ion. The products degraded by this base were found to contain triple bonds, and oxidized rapidly when exposed to air. In some cases they ignited spontaneously or exploded when shocked suggesting that the carbon was present in a highly reactive form.

As the above discussion brings out, PVDC can be dehydrochlorinated by almost any strong base. But in many cases, the bases used are also good nucleophiles. Therdfore, elimination and substitution occur competitively and complicate the interpretation of kinetic data.

Laurent and coworkers[40] neatly avoided this problem by using the LiCl/DMF system. Chloride ion is a strong base in DMF and in displacement reactions on chlorinated polymers, it does not change chemical composition. Therefore, the elimination process can be conveniently isolated. The decomposition was studied at 80 °C;

PVDC is only slightly soluble in DMF at this temperature and becomes even less soluble as it degrades. Brown degradation products showing an absorption maximum around 450-490 mμ were obtained. This material appears to be made up of short sequences of conjugated double bonds. Further reaction does not increase the sequence length. This reaction also goes beyond 50% elimination. The authors suggested that this was due to the formation of triple bonds, but could give no direct evidence in support of this conjecture. Nevertheless, this possibility is consistent with the probable mechanism of base induced decomposition.

There is no doubt that this is an ionic process. The evidence favors a concerted E2 elimination as the primary step in the process

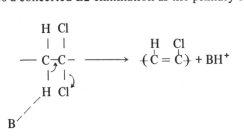

Laurent *et al.* feel that this is the rate controlling step and explains why PVDC and chlorinated PVC are more rapidly degraded than PVC.

The formation of the first double bond in a chain creates allylic activation of adjacent groups.

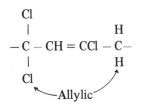

Substituents on an allylic carbon are much more reactive in ionic reactions than substituents in aliphatic carbon.[41] Therefore, the formation of one double bond makes subsequent elimination easier.

The direction in which the elimination reaction propagates depends on the mechanism involved. The chlorine substituent could eliminate by an E1 mechanism. (It is positioned on an allylic activated tertiary carbon which should favor ionization.)

$$
\begin{array}{c}
\underset{\underset{H}{|}}{\overset{\overset{H}{|}}{-C}}-\underset{\underset{Cl}{|}}{\overset{\overset{Cl}{|}}{C}}-\underset{}{\overset{\overset{H}{|}}{C}}=\underset{}{\overset{\overset{Cl}{|}}{C}}-\quad\xrightarrow{\ \ \text{Slow}\ \ }\quad \underset{\underset{H}{|}}{\overset{\overset{H}{|}}{-C}}-\underset{\underset{H}{|}}{\overset{\overset{+}{}}{C}}-\underset{}{\overset{\overset{H}{|}}{C}}=\underset{}{\overset{\overset{Cl}{|}}{C}}-
\end{array}
$$

$$\downarrow \text{ Fast}$$

$$
\underset{}{\overset{\overset{H}{|}}{-C}}=\underset{}{\overset{\overset{Cl}{|}}{C}}-\underset{}{\overset{\overset{H}{|}}{C}}=\underset{}{\overset{\overset{Cl}{|}}{C}}-
$$

The rate of propagation of the conjugated sequence in this mechanism is dependent on base concentration only indirectly though the rate of formation of primary double bonds (initiation process).

If propagation takes place by an E2 elimination, the reaction will be accelerated by allylic activation of the methylene group.

$$
\underset{B^{-}}{\overset{}{\underset{}{\overset{\overset{H}{|}}{-C}}=\underset{}{\overset{\overset{Cl}{|}}{C}}-\underset{\underset{\overset{|}{}}{\overset{|}{}}}{CH}-\underset{\underset{Cl}{|}}{\overset{\overset{Cl}{|}}{C}}-}}\quad\longrightarrow\quad \underset{}{\overset{\overset{H}{|}}{-C}}=\underset{}{\overset{\overset{Cl}{|}}{C}}-\underset{}{\overset{\overset{H}{|}}{C}}=\underset{}{\overset{\overset{Cl}{|}}{C}}-
$$

Growth in this case will occur in the opposite sense from that of the E1 mechanism. The rate of propagation should be first order in base concentration.

Initiation can take place at several points along a polymer chain. Depending on the relative positions of the primary double bonds, the possibility exists that two sequences can grow in opposite directions on the same chain. When they meet, a group is isolated between conjugated sequences. In E1 elimination, this will be a dichloromethylene group.

$$
\underset{\underset{Cl}{|}}{\overset{\overset{Cl}{|}}{-C}}=CH-CCl=CH-C-CH=CH=CCl-CH=CCl-
$$

But in E2 elimination, a methylene group is trapped instead:

$$
\begin{array}{c}
\underset{|}{H}\ \underset{|}{Cl}\ \underset{|}{H}\ \underset{|}{Cl} \\
-C=C-C=C-CH_2-
\end{array}
\qquad
\begin{array}{c}
\underset{|}{Cl}\ \underset{|}{H}\ \underset{|}{Cl}\ \underset{|}{H} \\
-C=C-C=C-
\end{array}
$$

In either mechanism, a high rate of initiation favors short con-jugated sequences because of the directional nature of the elimination.

9.2.3　Degradation in solution

In studies of the degradation of crystalline PVDC, the heterogene-ity of the reaction greatly complicates the analysis of kinetic data. Therefore, it is somewhat surprising that so little has been done with solution degradation.

There is no great difficulty in preparing homogeneous reactions. PVDC dissolves readily in hot chlorinated aromatic solvents and in several polar aprotic solvents. The latter solvents make an ideal medium for studying ionic elimination reactions. The advantage of solution reactions is that the kinetics are independent of morpho-logy. The rate of dehydrochlorination will be determined solely by the intrinsic reactivity of the polymer and the basicity of the medium.

Matheson and Boyer[1] reported probably the first study of homogeneous Saran degradation. They treated solutions of a 15% ethyl acrylate copolymer in dioxane with various amounts of morpholine. The polymer discolored in proportion to the amount of base added. Absorbance in the 400-700 mμ range increased and shifted to longer wavelengths. This appeared to correlate with the formation of conjugated polyene sequences of increasing length.

PVDC degrades in strong solvents at an appreciable rate even without added base. Davies and Henheffer[42] followed the reaction in HMPA and TMSO by an NMR technique. The data fit a first order plot with an activation energy of $\sim$ 34 kcal/mole. Within experimental error, this is the same as activation energies for solid state thermal decomposition, but the absolute rate in these solvents was more than an order of magnitude higher.

Grant[43] made a more detailed study of the decomposition of PVDC in HMPA. He also investigated a number of other systems and found a good correlation between rate and solvent polarity.

The kinetic data obtained in HMPA strongly resemble the data obtained by solid state thermal decomposition, but absolute rates are much higher.

The addition of a stronger base, pyridine, to a solution of PVDC in N-methyl pyrrolidone increased the rate in proportion to the amount added. The reaction was also first order in PVDC concentration. These results are consistent with an E2 elimination mechanism.

Even in good solvents like HMPA, the reaction becomes heterogeneous or gells after a time. An increase in molecular weight or polymer concentration accelerated phase separation. An increase in temperature had the opposite effect. This suggested that crosslinking is the cause of gelation.

Grant was not able to characterize the degraded products completely. But he observed that they were crosslinked and discolored to an extent proportional to reaction time. The IR spectra resembled those of thermally degraded PVDC.

Though Grant assumed that gelation was due to crosslinking, he did not follow changes in molecular weight to prove this. Matheson and Boyer did however observe slight increases in molecular weight.

In a more quantitative study, Jackson and Reid[44] followed changes in intrinsic viscosity with time in several solvents. They found that $[n]$ dropped rapidly in polar solvents but changed very little in o-dichlorobenzene. This work suggests that chain scission must be taking place.

Recent investigations have shown that the rate of degradation of PVDC is very sensitive to solvent structure.[45] The rate is lowest in non-polar solvents like tetrahydronaphthalene; it is substantially higher even in weakly polar solvents like benzyl benzoate though still orders of magnitude lower than the rates observed by Grant.

The degradation in a non-polar solvent like o-dichlorobenzene apparently involves a free radical propagation step.[7] The rate of dehydrochlorination is much lower than that for a solid state reaction. The activation energy, however, is about the same.

The general similarity in the dehydrochlorination kinetic pattern and the constancy of the activation energies imply that some common feature exists in the mechanisms of these reactions. As suggested in the case of solid state reactions, this is probably a common initiation process involving ionic elimination to form primary double bonds.

The mode of propagation, however, depends on the nature of the reaction medium. Unzipping via a free radical chain reaction seems to occur in o-dichlorobenzene, but some form of sequential ionic elimination must take place in HMPA.

A major difference between homogeneous and heterogeneous dehydrochlorination is that initiation can take place at random in the former but only on the crystal surface in the latter. The frequency of initiation limits the length of conjugated sequences in either case but morphology probably also has incluence on the solid state case.

A major difference between free radical and ionic dehydro-chlorinations is in the total amount of chlorine that can be removed from the polymer. Typically, thermal degradation will become very slow beyond ~ 50% elimination. The ionic reaction can proceed much further up to virtually 100% in the presence of strong bases. This is probably associated with the fact that the second chlorine can be eliminated ionically but is relatively stable to radical attack.

The autocatalytic character of these reactions is generally ascribed to allylic activation. But another explanation is that either HCl or Cl⁻ (depending on medium) catalyze the reaction. It is known that Cl⁻ acts as a base in polar aprotic solvents. Since it is formed in the degradation process, the concentration of "base" would actually increase with time. This condition alone could be responsible for the autocatalytic behavior.

9.2.4 Acid catalyzed degradation

The apparent acceleration of the decomposition of PVDC in the presence of proton acids especially HCl was observed by the early Dow workers. This was interpreted by some as direct acid catalysis.[2] However, dry HCl was shown to have no effect on the rate of decomposition of pure PVDC.[9] Its effect was later shown to be indirect, the acid acting to convert metal oxides, present in the polymer as impurities, to catalytic chloride salts.[46] Recent studies have confirmed this mechanism.[45]

The transition metal chlorides, especially $AlCl_3$, $FeCl_3$ and $ZnCl_2$ are potent dehydrochlorination catalysts.[47] For example, only 7.3×10^{-3} mole % $FeCl_3$ adsorbed onto PVDC powder will cause the polymer to release ~ 10 meq HCl/gram (about 50% decomposition) in less than 10 minutes at 100 °C *in vacuo*. The reaction is so fast that the powder virtually explodes.

The products at 50% dehydrochlorination are soluble low molecular weight polychloroacetylenes. The reduction in molecular weight indicates that both dehydrochlorination and chain cleavage occur in the presence of Lewis acids. The catalytic activity seems to correlate with the activity of salts as catalysts in Friedel Crafts reactions. This suggests that a carbonium ion mechanism is involved.

Lewis acid catalysis explains the difficulties encountered in trying to fabricate Saran resins.[48] The polymer, when heated, releases small amounts of HCl which attack the metal surface of the extruder. The metal chloride salt thus formed catalyzes further release of HCl. The process is extremely autocatalytic and has been known to lead to an explosion.

REFERENCES

1. L. A. Matheson and R. F. Boyer, *Ind. Eng. Chem.*, **44**, 867 (1952).
2. C. B. Havens, NBS No. 525, 107 (1953).
3. F. H. Winslow, W. O. Baker and Y. A. Yager, *Proc. First and Second Conf. Carbon*, p. 93 (1956).
4. L. G. Tokareva, N. V. Mikhailov and V. S. Klimenkov, *Russ. Colloid. J. (Eng. Trans.)*, **18**, 595 (1956).
5. J. R. Dacey and D. G. Thomas, *Trans. Faraday Soc.*, **50**, 740 (1954).
6. J. R. Dacey and D. A. Cadenhead, *Proc. Fourth Carbon Conf.*, p. 315 (1960).
7. G. M. Burnett, R. A. Haldon and J. N. Hay, *Eur. Polym. J.*, **3**, 449 (1967).
8. D. H. Davies, D. H. Everett and D. J. Taylor, *Trans. Faraday Soc.*, **67**, 382 (1971).
9. D. H. Everett and D. J. Taylor, *Trans. Faraday Soc.*, **67**, 402 (1971).
10. R. D. Bohme and R. A. Wessling, *J. Appl. Polym. Sci.*, **16**, 1761 (1972).
11. A. Bailey and D. H. Everett, *Nature*, **211**, 1082 (1966).
12. A. Bailey and D. H. Everett, *J. Polym. Sci. A-2*, **7**, 87 (1969).
13. R. A. Wessling, *J. Appl. Polym. Sci.*, **14**, 1531 (1970).

14. L. Mandelkern, *Crystallization of Polymers*, McGraw-Hill, N.Y. (1964).
15. J. R. Dacey and R. G. Barradas, *Can. J. Chem.*, **41**, 180 (1963).
16. R. F. Boyer, *J. Phys. Coll. Chem.*, **51**, 80 (1947).
17. B. Dolezel, M. Pegorara and E. Beati, *Eur. Polym. J.*, **6**, 1411 (1970).
18. J. N. Hay, *J. Polym. Sci. A-1*, **8**, 1201 (1970).
19. I. R. Harrison and E. Baer, *J. Coll. Interface Sci.*, **31**, 176 (1969).
20. M. Onozuka and M. Asahina, *J. Macromol. Chem.*, C3, 235 (1969).
21. C. A. Brighton, *Brit. Plastics*, **20**, 62 (1955).
22. N. L. Zutty and F. J. Welch, *J. Polym. Sci. A*, **1**, 2289 (1963).
23. G. N. Burnett, R. A. Halden and J. N. Hay, *Eur. Polym. J.*, **4**, 83 (1968).
24. R. A. Haldon and J. N. Hay, *J. Polym. Sci. A-1*, **6**, 951 (1968).
25. N. Murayama and Y. Amagi, *Polym. Lett.*, **4**, 115 (1966).
26. L. K. Frevel, The Dow Chemical Company, unpublished results.
27. P. J. Flory, *Principles of Polymer Chemistry*, Cornell Univ. Press, Ithaca, N.Y. (1953).
28. D. Dollimore and R. R. Heal, *Carbon*, **5**, 65 (1967).
29. B. V. Lebedev, I. B. Rabinovich and V. A. Budarina, *Polym. Sci. USSR*, **9**, 545 (1967).
30. F. A. Bovey, *The Effects of Ionizing Radiation on Natural and Synthetic Polymers*, Interscience, N.Y. (1958).
31. G. Oster, G. K. Oster and M. Kryszewski, *J. Polym. Sci.*, **57**, 937 (1962).
32. G. Oster et al., *Nature*, **191**, 164 (1961).
33. M. Kryszewski and M. Mucha, *Bull. Acad. Polon. Sci. (Ser. Sci. Chim)*, **13**, 54 (1965).
34. D. E. Harmer and J. A. Raab, *J. Polym. Sci.*, **55**, 821 (1961).
35. R. A. Wessling, presented at Fall ACS Meeting, 1976.
36. T. Nakagawa, *Kogyo Kagaku Zasshi*, **71**, 1272 (1968).
37. S. S. Barton, J. R. Dacey and B. H. Harrison, *Org. Coating and Plastics Preprint ACS*, **31**, 768 (1971).
38. E. Tsuchida, C. Shih and J. Shinohara, *Polym. Lett.*, **3**, 643 (1965).
39. E. Tsuchida, C. Shih, I. Shinohara and S. Kambere, *J. Polym. Sci. A*, **2**, 3347 (1964).
40. G. Laurent, J. P. Roth, P. Kempp and J. Parrod, *Bull. Soc. Chim. France*, 2923 (1966).
41. C. K. Ingold, *Structure and Mechanism in Organic Chemistry*, Cornell Univ. Press, Ithaca, N.Y. (1953).
42. D. H. Davies and P. M. Henheffer, *Trans. Faraday Soc.*, **66**, 2329 (1970).
43. D. H. Grant, *Polymer*, **11**, 581 (1970).
44. D. L. C. Jackson and W. S. Reid, *Nature*, **162**, 29 (1948).
45. D. E. Agostini and A. L. Gatzke, *J. Polym. Sci. Chem.*, **11**, 649 (1973).
46. R. A. Wessling, unpublished results.
47. R. D. Bohme and R. A. Wessling, U.S. 3,852,223, to The Dow Chemical

Company (1974).
48.   R. A. Wessling and F. G. Edwards, *Enc. Polym. Sci. Tech.*, **14**, 540 (1971).

CHAPTER 10

Carbonization of Saran

Saran carbon is the name usually applied to carbonaceous products derived from the pyrolysis of polyvinylidene chloride or Saran resins. The materials obtained by this procedure fall into the general category of non-graphitizing carbons.[1] Their structures can crudely be described as an assembly of randomly oriented graphite microcrystals in an amorphous carbon matrix. Saran carbon can be further characterized as having a microporous structure. The particular structure developed is very dependent on preparative conditions.

PVDC can be carbonized merely by heating to 700 °C in an inert atmosphere. The thermally induced dehydrochlorination reaction gives a quantitative yield of carbon and HCl gas provided the heating is done slowly. The resulting carbon is free of all impurities save a trace of chlorine. This gives Saran carbon some distinct advantages over carbons obtained by the pyrolysis of other organic materials such as cellulosics.

Saran carbons contain very small pores, which give them molecular sieve properties. Both porosity and pore size can be selected to some extent by varying the techniques used to make the carbon. Samples with pores as small as 6 Å and surface areas of 1400 m^2/g have been reported. These carbons have a high adsorbing capacity even when unactivated.

Saran carbon dates back to at least 1935.[2] It was first prepared accidentally by Dow researchers attempting to heat fabricate PVDC.

The first literature reference to Saran carbon as an adsorbent appeared in 1948 when Emmett[3] mentioned work done during World War II with Saran carbon. Then in 1949 Pierce et al.[4,5]

reported a study of adsorption on Saran carbon. At about the same time, an analysis of its structure by Franklin[6] appeared.

In 1954 Dacey and Thomas[7] pointed out that Saran carbon had molecular sieve characteristics. Dacey continued the study of its preparation, properties and surface chemistry throughout the following decade.[8-15] In that same period, a number of papers were published on the preparation and properties of Saran carbon,[16-27] and the subject was discussed at several conferences on carbon.[28-30] But the relation between the starting polymer morphology and the properties of the resultant carbon was not appreciated.

The significance of morphology was first pointed out by Everett and Bailey in 1966.[31] They expanded upon this concept in a later paper[32] showing that there was a direct relation between the lamellar crystal morphology found in PVDC and the carbon platelets obtained by controlled degradation. This is currently an active area of research for Everett and coworkers.[33-35]

Chemical methods of carbonizing Saran have received relatively little attention. Evans and Flood[36,37] and Tsuchida et al.,[38,39] described the use of potassium amide in liquid ammonia to effect dehydrochlorination. The resulting carbons had interesting chracteristics but apparently this approach was not pursued further.

Barton et al.[15] used alcoholic KOH, to induce chemical carbonization. In this case, the polymer contained substantial amounts of chemically bound oxygen. The pore size was much smaller than that of thermally degraded polymer.

10.1 CHEMISTRY OF SARAN CARBON

10.1.1 Preparative methods

There are two basic techniques for carbonizing PVDC thermally. Although both result in essentially the same product, the morphology is considerably different. In one approach, favored by Dacey[11] and Everett,[31] PVDC is first degraded below its melting point to an infusible char; the latter is then heated further at high temperatures to complete dehydrochlorination. This process is based on the recognition that solid state degradation gives more control over the surface area and pore size of the resultant carbon.

The other method involves heating at a fairly high rate until a temperature is reached where dehydrochlorination is complete.[40,41] Using this technique, PVDC first melts before any significant degradation occurs. When HCl gas begins to evolve, it causes the melt to foam. The foam gradually crosslinks and hardens as the reaction proceeds. The carbon resulting from the process contains large macropores.

A better understanding of these processes can be gained by reviewing briefly the mechanism of the dehydrochlorination reaction. At least four basic reactions are involved.[11,19,23]

1) Primary dehydrochlorination to polychloroacetylene:

$$\left(CH_2-CCl_2\right)_n \xrightarrow{\Delta} \left(CH=CCl\right)_n + nHCl\uparrow$$

2) Diels-Adler condensation between conjugated sequences:

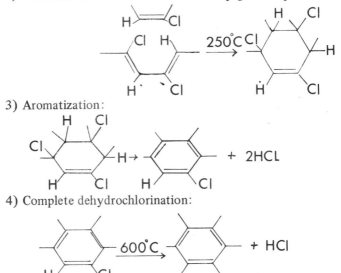

3) Aromatization:

4) Complete dehydrochlorination:

According to this mechanism, approximately half of the available HCl is lost in Reaction 1, about 1/3 in Reaction 3 and the final 1/6 in Reaction 4. Reaction 1 takes place in the solid state between 130-190 °C. A small amount of 2 and 3 probably occur

at the high end of this range. But, if the temperature is not
increased, the dehydrochlorination rate eventually drops to zero.
The total amount eliminated depends on the temperature but does
not exceed ~ 60% even at 190 °C.

Temperatures of 250-350 °C are required to make the Diels-
Alder reaction go at a rapid rate. The initial ring formation shown
in Reaction 2 is probably followed immediately by Reaction 3.
The tendency to aromatize at these temperatures should be very
strong.

The temperature must be raised even further to remove the
remaining chlorine. This is usually complete in the range of 600-700
°C. Residual chlorines are probably substituted on vinylic or aromatic carbons; hence, the high temperatures needed to complete
the carbonization. These changes are clearly illustrated in the stepwise degradation experiments of Winslow *et al.* Their results are
shown in Figure 10.1.

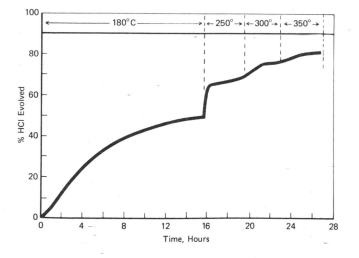

FIGURE 10.1 Stepwise pyrolysis of polyvinylidene chloride (Ref. 19).

Looking back now to the processes used to carbonize PVDC, we see that the first is designed to carry out the initial reaction before aromatization. The rapid heating process, on the other hand, allows initial dehydrochlorination and aromatization to occur simultaneously.

Actually, the extent of dehydrochlorination below 190 °C necessary to freeze in the structure is only about 10-20%. Once this level is reached, the char can be heated rapidly to 700 °C without altering the morphology. Rapid heating to temperatures in excess of 350 °C without prior decomposition reduces the yield of carbon, and quantities of VDC monomer and 1,3,5-trichlorobenzene are produced instead.[42] At these high temperatures, chain scission and depolymerization become competitive with dehydrochlorination and crosslinking. The technique of rapid pyrolysis is not a suitable way of making carbon.

A lack of appreciation of the importance of heating rate has led to some confusion about the nature of the carbonization reaction. A correlation has been established between the nature of the initial decomposition of char forming polymers and the ability to graphitize on subsequent heating.[1] Polymers, like PVC, that pass through a soft (or plastic) stage during dehydrochlorination form carbons that graphitize readily. Polymers that decompose without softening form non-graphitizing carbons.

PVDC produces carbons of the latter variety because it crosslinks to a rigid structure early in the decomposition. This has been amply documented by many workers. But, from time to time, researchers have deduced from morphological examinations that it must have passed through a plastic stage, but nonetheless is still non-graphitizing.[41]

These cases involved a constant heating rate technique such as a DTA or TGA experiment at 5-20 °C/min. At these rates, the polymer melts and foams before carbonization takes place.

Constant heating rate experiments such as DTA seem ill-suited for studying reactions such as the dehydrochlorination of PVDC. The various reactions that occur have induction periods and different activation energies. DTA experiments do not bring out these details, and, consequently, could lead to erroneous conclusions.

10.1.2   Structure

A structure for Saran carbon was proposed by Franklin based on an analysis of x-ray diffraction data. In a following paper, she explored the effect of thermal treatment on the structure and detailed the differences between graphitizing and non-graphitizing carbons.[1,6] While her major conclusions have been confirmed by later studies[11,19,20,31,32,40] recent work suggests that the proposed mechanism of graphitization needs revision.[43]

Franklin studies a carbon obtained by pyrolyzing PVDC at 600 °C. It was then heated an additional two hours at 1000 °C. Her analysis showed that 35% of the carbon in this sample was in an essentially amorphous state. However, the remaining 65% was organized into graphite microcrystals (or crystallites). These are small ordered regions averaging 16 Å in diameter containing about two graphite layers. The bond lengths in the layers were normal for a graphite structure, 1.42 Å. The interlayer distance, however, was 3.7 Å; significantly larger than the 3.19 Å spacing in graphite. This difference is the result of the small number of layers in the crystallites contained in Saran carbon.

The average distance between the ordered regions was found to be 26 Å. Franklin was unable to describe the exact nature of the interparticle regions. The amorphous carbon atoms are distributed within them, but their exact disposition is uncertain. They are probably bonded to the free valences on the edges of a graphite layer and serve to knit the entire mass together into a tightly crosslinked hard solid. A comparison of apparent density to true density shown in Figure 10.2 indicates that most of the inter-crystalline region is made up of micropores, with a diameter of 10 Å.

The differences between non-graphitizing carbons like Saran carbon and graphitizing carbons like PVC carbon are minor unless the specimens have been heated above 1000 °C. The low temperature graphitizing carbons contain a similar graphite micro-crystalline turbostratic structure but with an average of four layers per crystallite rather than two. In addition, they lack the microporous structure of the non-graphitizing carbons and are relatively soft suggesting a less crosslinked structure.

Major differences between the two types of carbon become

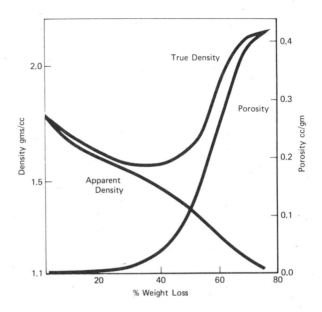

FIGURE 10.2 The changes in density and porosity as decomposition proceeds (Ref. 10).

apparent with additional thermal treatment. In both cases, the amorphous component decreases in concentration with a resultant growth in the diameter of the graphite layers. No change in the number of layers is observed until the amorphous phase disappears.

Following these changes, the graphite crystallites begin to grow both in diameter, and by addition of new layers. In non-graphitizing carbons, crystallites with diameters of ~ 70 Å and containing about 12 layers can be achieved but they do not grow further even by prolonged heating at 3000 °C. The crystallites are randomly oriented and are of the same dimension as the primary carbon particles indicating that further growth across void spaces does not occur.

When graphitizing carbons are heated in the same way, the

graphite crystallites continue to grow to macroscopic dimensions. Growth by layer addition is faster than lateral growth. There is relatively little fine pore structure in these carbons and a greater degree of parallel orientation of the primary crystallites. Both factors promote the growth of a continuous graphite phase.

Franklin has proposed a mechanism of growth to account for non-graphitization. The only model which can rationalize her observations is one in which growth of the graphite crystals takes place by the addition of layers to an existing crystal. If growth were the result of diffusion of individual carbon atoms to the crystal surface, micropores would be no barrier to growth. In fact, the microcrystals on non-graphitizing carbons do not grow across particle boundaries. One would expect, on the other hand, that micropores would present a serious obstacle to the movement of a layer of carbon atoms.

The magnetic properties of Saran and PVC carbons were studied by Blayden and Wescott[21] with the idea of showing differences between graphitizing and non-graphitizing carbons. They also examined structure and confirmed the results reported by Franklin. At low temperatures, both polymers had the characteristic turbostratic structure but on further heating, PVC carbon was converted to graphite while Saran carbon retained its turbostratic structure. Polyacrylonitrile carbon was reported to behave much like Saran carbon.

Winslow and coworkers[18,19] in a comparative study of various polymer derived carbons also made an x-ray diffraction analysis of Saran carbon. They used a stepwise heating technique and examined structure at intermediate states of carbonization. The polyene products obtained by pyrolysis below 300 °C had typical amorphous structures. The powder patterns were similar to those obtained from crosslinked polymers. There was no evidence of ordered structures. Further heating to 800 °C completely carbonized the Saran sample with no development of a graphite structure. They point out that this is consistent with the hypothesis that Saran carbon develops a homogeneous network of aromatic crosslinks below 300 °C that freezes in the structure. ESR studies are also in agreement with this model.[18,44]

A number of other studies have been made on the structure of

PVDC carbon and its non-graphitizing characteristics. Dubinin[25] has summarized the work of Russian investigators. Gilbert and co-workers[24] made a comparison of various carbonized polymers by TGA methods. Dollimore and Heal[40] carried out a similar study using DTA and gas evolution as well as TGA. They found that non-graphitizing carbons were formed from polymers that showed a large exothermic reaction at low temperatures. Examples were PVDC, polyacrylonitrile and various cellulosic polymers.

10.1.3 Morphology

There are two aspects of morphology to consider: the macro-scopic shape of the carbon particles and the microporous structure. Both are strongly influenced by preparative conditions. Subsequent heat treatments and activation processes may also affect morphology, particularly microporosity.

Everett and Bailey[31,32] have shown that when PVDC is degraded below its melting point to about 10% HCl loss, the particles become completely crosslinked. The morphology existing at that point becomes "frozen in" and persists in the final carbon obtained after complete dehydrochlorination at high temperatures. Since the primary PVDC particles exist in a lamellar structure, a similar morphology is shown by Saran carbon. The large pores in the carbon are void spaces between the lamellar particles.

Fine pore structure develops by continued dehydrochlorination of the crosslinked polymer particles. The loss of HCl without com-pensating shrinkage must produce cracks in the particle perpendicu-lar to the lamellar plane. As a consequence, the cracks must be very small and have a slit-like morphology. The carbon particles appear to be 200-500 Å thick whereas the original polymer crystals were 90-100 Å. Everett and Bailey suggest that the greater thickness of the carbon particles is due to crosslinking between lamellae. But, at the temperatures where they carried out the preliminary degradation, the polymer crystals thicken by an annealing process. They may simply have grown to 200-500 Å thickness before cross-linking.

Everett and Bailey point out that when PVDC is carbonized above its melting point, the carbon has a much more uniform structure. Micrographs published recently by Boult et al.[41] show that it is essentially an open cell foam. Such carbons have a high degree of macroporosity and do not show adsorption hysteresis as do carbons prepared by initial solid state degradation.

The pore sizes in Saran carbons have been deduced indirectly from various gas adsorption, density and surface area measurements. Dacey[7] tried to deduce pore size by studying adsorption of molecules with various cross sections. He observed a significant drop in apparent porosity calculated from tetraethylmethane suggesting that pores were slightly larger than the cross section of this molecule. This is consistent with Franklin's estimate of 10 Å.

In a later study, Dacey[12] estimated pore size from helium adsorption measurements and confirmed this estimate. His data indicated that much of the surface area was contained in slit-like pores < 9 Å in diameter. Some of the pores fell in the range of 9-15 Å and the remainder were still larger.

Dubinin[25] also proposed the existence of three types of pores which he termed micro-, transitional and macro-pores. These results are consistent both with Franklin's structure and the more detailed model prepared by Everett.

Many workers including the Russian group have observed that the apparent porosity of Saran carbon reaches a maximum when thermally activated at ~ 1300 °C. Surface areas approaching 1400 m^2/g are achieved at this temperature.[26,45] Further increases in temperature, however, produce a rapid drop in porosity, as shown in Figure 10.3. This appears to be due to a sealing off of pores at the carbon particle surface as the amorphous carbon component undergoes crystallization.

10.2 PROPERTIES AND USES

10.2.1 Adsorptive properties

Most of the interest in Saran carbon relates to its adsorptive properties. Compared to other polymer derived carbons, it has a high capacity without activation. Kipling and Wilson[22] compared the adsorptive capacity of 18 carbons to various vapors and found that none of these could approach the performance of Saran carbon unless activiated.

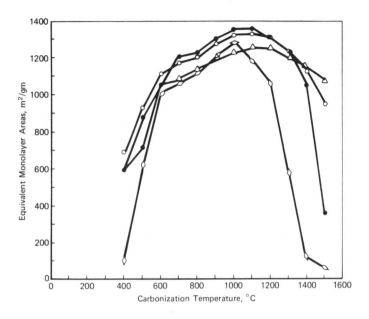

FIGURE 10.3 Surface areas of polyvinylidene carbons (Ref. 44).
$\Diamond\,CO_2$; $\triangle$ Neopentane; $\bullet$ Isobutane; $\circ$ Butane.

The actual capacity varies both with starting material (i.e., whether Saran B resin or PVDC) and thermal treatment. The values given by Dacey[11] are typical of what has been observed. These data are shown in Table 10.1.

The selectivity of these carbons is also evident in Dacey's data. As the molecular size increases above that of tetraethyl methane, the capacity drops to zero.

Kaiser[46] has reported that PVDC carbons can act as effective chromatographic substrates. They yield good separation of low molecular weight species in gas, thin layer and column chromatography. The contaminant free, hydrophobic surfaces characteristic of these carbons cut down the tailing usually observed with polar molecules. Water, for example, comes off in a sharp peak before

TABLE 10.1

Saturation Values of Various Adsorbates on Samples of
Completely Pyrolyzed Saran (Ref. 11)

Adsorbate	cm^3/g	Adsorbate	cm^3/g
Nitrogen	0.43	Naphthalene	0.37
n-Pentane	0.42	Neopentane	0.34
Benzene	0.42	Cyclohexane	0.34
Toluene	0.43	Tetraethylmethane	0.27[a]
cis-Butene-2	0.43	α-Pinene	Negligible
trans-Butene-2	0.43		
Argon	0.41		

[a] Non-equilibrium value after 6 days.

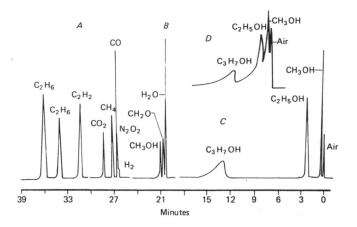

FIGURE 10.4 A: Carbon molecular sieve "B", 80/100 mesh, column 1 m length, 2 mm diam., TPGC from 40 to 180 °C, heating rate 10 °C/min.; Hc; TC-detector. B: Same as A, but isothermal analysis, 200 °C. C: Same as A, but isothermal analysis, 250 °C. D: Activated charcoal, same column dimensions as A, same conditions as C (Ref. 46).

methane as shown in Figure 10.4. Hydrocarbons are strongly held and as molecular weight increases cannot be removed without decomposition.

The above descriptions suggest the use of Saran carbon in separation, purification and analysis applications. Walker[47] has shown them to be more effective than conventional activated charcoal in adsorbing krypton at the low pressures found in atomic reactors. Kaiser[48] has shown that water can be determined quantitatively in a Saran carbon packed VPC. Zlatkis *et al.*[49] have described the use of a column of the same type to analyze for trace impurities in ethylene. The success of these ventures suggests that this area of application will grow in the future.

10.2.2 Other properties

The electrical and magnetic properties of PVDC carbon have been studied by several groups including Winslow *et al.*[18] Blayden and Wescott[21] and more recently Dacey and coworkers.[14,15] Completely carbonized PVDC is a fairly good conductor with resistivity = 0.01 ohm-cm at 25 °C. The partially carbonized chars are also semiconducting with resistivity falling from 10^{16} ohm-cm to 10^{-2} as dehydrochlorination increases from 50 to 100%.

The magnetic susceptibility increases with pyrolysis temperature up to the maximum studied (3000 °C)[21] but never attains the value of graphite or graphitizing carbons. The latter is $- 5.84 \times 10^6$ cgs units/g at room temperature; for PVDC carbon heated to 3000 °C, the maximum value was $- 4.65 \times 10^6$. These differences correlate with the structure proposed for non-graphitizing carbons.

The electrical properties of Saran carbon are altered by adsorption. Dacey *et al.* have shown that adsorption of H bonding liquids like H_2O increases resistivity while causing the carbon to shrink. Adsorbed aprotic gases cause the resisitivity to fall. This can be explained by assuming that H bonds simultaneously pull graphite layers closer together and tie up charge carriers.

10.2.3 Material properties

The use of Saran carbon as a structural material has received little attention. The hardness and abrasion resistance of these carbons have been pointed out by several workers, but the high internal porosity tends to make them weak mechanically. This is clearly evident in the attempts by Boucher et al.[50] to make high strength carbon fibers from Saran fiber.

The found that their fibers could be converted to carbon without loss of fiber geometry by predegrading the surface with alcoholic KOH and then heating in a programmed manner. The polymer is first dehydrochlorinated at 140 °C until 50% reaction, then to 70% reaction at 180 °C and finally to 850 °C at 50 °/h.

At this point, the fibers show evidence for graphite microcrystal formation and exhibit surface areas of ~ 1000 m^2/g. The internal porosity is retained even after treatment of 2500 °C, but pore entrances are sealed off. The mechanical properties are greatly inferior to those of carbon fibers obtained from polyacrylonitrite fiber.

REFERENCES

1. R. E. Franklin, *Proc. R. Soc.*, **A209**, 196 (1965).
2. P. Banner, The Dow Chemical Company, private communication.
3. P. H. Emmett, *Chem. Rev.*, **43**, 69 (1948).
4. C. Pierce, J. W. Wiley and R. N. Smith, *J. Phys. Coll. Chem.*, **53**, 669 (1949).
5. C. Pierce, R. N. Smith, J. W. Wiley and H. Cordes, *J. Am. Chem. Soc.*, **73**, 4551 (1951).
6. R. E. Franklin, *Acta Crystallogr.*, **3**, 107 (1950).
7. J. R. Dacey and D. G. Thomas, *Trans. Faraday Soc.*, **50**, 740 (1954).
8. J. R. Dacey and D. G. Thomas, *Can. J. Chem.*, **33**, 344 (1955).
9. J. R. Dacey, J. C. Clunie and D. G. Thomas, *Trans. Faraday Soc.*, **54**, 250 (1958).
10. J. R. Dacey and D. A. Cadenhead, *Proc. 4th Conf. Carbon*, Pergammon Press, N.Y., p. 315 (1960).
11. J. R. Dacey and R. G. Barradas, *Can. J. Chem.*, **41**, 180 (1963).
12. J. R. Dacey and M. H. Edwards, *Can. J. Phys.*, **42**, 241 (1963).
13. J. R. Dacey, G. J. C. Frohnsdorff and J. T. Gallagher, *Carbon*, **2**, 41 (1964).
14. J. R. Dacey, D. F. Quinn and J. T. Gallagher, *Carbon*, **4**, 73 (1966).
15. S. S. Barton, J. R. Dacey and B. H. Harrison, *Org. Coatings Plastic Preprints*, **31**, 786 (1971).
16. R. V. Culver and N. S. Heath, *Trans. Faraday Soc.*, **51**, 1569 (1955).
17. N. S. Heath and R. V. Culver, *Trans. Faraday Soc.*, **51**, 1575 (1955).
18. F. H. Winslow, W. O. Baker and W. A. Yager, *J. Am. Chem. Soc.*, **77**,

4751 (1955).

19. F. H. Winslow, *et al.*, *Proc. 1st and 2nd Conf. Carbon*, University of Buffalo Press, Buffalo, N.Y., p. 93 (1956).

20. D. H. Everett and E. Redman, *Fuel*, **42**, 219 (1963).

21. H. E. Blayden and D. T. Westcott, *Proc. 5th Conf. Carbon*, Pergammon Press, N.Y., p. 97 (1962).

22. J. J. Kipling and R. B. Wilson, *Trans. Faraday Soc.*, **56**, 557 (1960).

23. J. J. Kipling and R. B. Wilson, *Trans. Faraday Soc.*, **56**, 562 (1960).

24. J. B. Gilbert, J. J. Kipling, B. McEnaney and J. N. Sherwood, *Polymer*, **3**, 1 (1962).

25. M. M. Dubinin, *Proc. 5th Conf. Carbon*, Pergammon Press, N.Y., p. 81 (1962).

26. H. Marsh and W. F. K. Wynne-Jones, *Carbon*, **1**, 269 (1964).

27. T. G. Lamond and H. Marsh, *Carbon*, **1**, 293 (1964).

28. *Conference on Industrial Carbon and Graphite*, S.C.I. London (1958).

29. *Tenth Symposium of Colston Research Society*, London (1958).

30. *Second Conference on Industrial Carbon and Graphite*, S.C.I., London (1965).

31. A. Bailey and D. H. Everett, *Nature*, **211**, 1082 (1966).

32. A. Bailey and D. H. Everett, *J. Polym. Sci. A-2*, **7**, 87 (1969).

33. D. H. Davies, D. H. Everett and D. J. Taylor, *Trans. Faraday Soc.*, **67**, 382 (1971).

34. D. H. Everett and D. J. Taylor, *Trans. Faraday Soc.*, **67**, 402 (1971).

35. L. B. Adams, E. A. Boucher and D. H. Everett, Paper prepsented at *8th Conf. Carbon*, Boston (1969).

36. B. Evans and E. A. Flood, *Can. J. Chem.*, **45**, 1713 (1967).

37. B. Evans and E. A. Flood, U.S. 3,516,791 (1970).

38. E. Tsuchida, C. Shih, I. Shinohara and S. Kambara, *J. Polym. Sci. A*, **2**, 3347 (1964).

39. E. Tsuchida, C. Shih and I. Shinohara, *Polym. Lett.*, **3**, 643 (1965).

40. D. Dollimore and G. R. Heal, *Carbon*, **5**, 65 (1967).

41. E. H. Boult, H. G. Campbell and H. Marsh, *Carbon*, **7**, 700 (1969).

42. S. Tsuge, T. Okumoto and T. Takeuchi, *Makromol. Chem.*, **123**, 123 (1969).

43. J. Maire and J. Mering, in P. L. Walker (ed.), *Chemistry and Physics of Carbon*, Vol. 6, M. Dekker, N.Y., p. 125 (1970).

44. D. Campbell, C. Jackson, H. Marsh and W. F. K. Wynne-Jones, *Carbon*, **4**, 159 (1966).

45. T. G. Lamond, J. E. Metcalfe and P. L. Walker, *Carbon*, **3**, 59 (1965).

46. R. Kaiser, *Chromatographia*, **3**, 38 (1970).

47. O. P. Mahajan and P. L. Walker, *J. Coll. Interface Sci.*, **29**, 129 (1969).

48. R. Kaiser, *Chromotographia*, **2**, 453 (1969).

49. A. Zlatkis, H. R. Kaufman and D. E. Durbin, *J. Chromatog. Sci.*, **8**, 416 (1970).

50. E. A. Boucher, R. N. Cooper and D. H. Everett, *Carbon*, **8**, 597 (1970).

CHAPTER 11

Technology

Commercial products include a variety of polymers containing from 60 to 95% VDC.[1,2] The properties of a particular material depend on the kind and amount of comonomers and on the method of polymerization. Saran resins are often further modified by the incorporation of plasticizers. The combination of co-polymerization and plasticization permits the design of polymers ranging from soft rubbery materials to hard, rigid plastics.

The major applications for Saran lie in the field of packaging where its low permeability and good optical properties are of value. The traditional use of Saran resins as thermoplastics for molding is no longer as important. But it is still used in a number of applications that require excellent chemical resistance or non-flammability combined with low cost.

As mentioned in Chapter 1, the more important systems are those containing high proportions of VDC with varying amounts of VC, acrylonitrile, acrylates and methacrylates either alone or in combination. Very often, another comonomer with specific functional properties is included in small amounts. Gabbett and Smith[2] have compiled a list of terpolymer systems described in the patent art.

Saran resins are often formulated with plasticizers to get better low temperature flexibility. The more common types include alkyl esters of adipic, sebasic and citric acids. Polymeric plasticizers such as polyethylene-co-vinylacetate and polyurethane rubbers are also used. Most of the common PVC plasticizers can be used with Saran as well. But compatibility with high VDC resins is poor.

11.1 STABILIZATION

Saran polymers are normally stabilized to avoid serious decomposition during fabrication and use. As the description of its degradation chemistry implies, the stabilization of Saran is a formidable problem. Nonetheless, the art of stabilization has become highly developed through mostly on an empirical basis. The practical aspects of stabilizing chlorinated polymers has been described in the treatise by Chevassus and De Broutelles.[3] They also include an extensive compilation of patents on the subject. This literature concerns mainly PVC but many, though not all, PVC stabilizing systems can be used with Saran as well.

A given stabilizer recipe may have several ingredients. Many of the individual components by themselves and in excess will catalyze the degradation of the polymer. Therefore, the components must be finely balanced. These recipes are described mainly in the patent literature, and very little is known about specific mechanisms. In many cases, combinations of various additives have a synergistic effect.

The stabilizing system has to protect the polymer from unavoidable exposure to heat and light and contact with metals. It must also scavenge HCl gas and minimize the discoloration of the polymer. The latter problem is particularly difficult because severe discoloration can be generated by very little degradation.

The characteristics desired in a Saran stabilizer system have recently been summarized.[4] This list is reproduced below:

1. adsorb, or combine, with hydrochloric acid gas in an irreversible manner under the conditions of use, but not have such strong affinity as to strip HCl from the polymer chain.

2. act as a selective ultraviolet light absorber to reduce the total ultraviolet energy absorbed in the polymer,

3. contain a reactive dienophilic molecule capable of destroying the discoloration by reacting with, and breaking up, the color-producing, conjugated polyene sequences,

4. possess anti-oxidant activity in order to prolong the induction period of the oxidation process and prevent the formation of carbonyl groups and other chlorine labilizing structures resulting from oxidation of polymer molecules,

5. have the ability to chelate metals, such as iron, and prevent the forma-
    tion of metallic chloride which acts as a catalyst for polymer degradation.

Many stabilizers show more than one function. Epoxides, for
example, are believed to act both as HCl acceptors and chelating
agents. They may also have some antioxidant activity. Probably
for this reason, epoxy compounds such as epoxidized soya bean oil
have become important commercial stabilizers.

Other acid acceptors commonly used include magnesium oxide,
tetrasodium pyrophosphate, metal salts of fatty carboxylic acids
and organo-tin compounds. The latter are used mostly in PVC where
they prevent discoloration as well.

Light stabilizers for Saran are commonly based on derivatives
of salicyclic acid and benzophenone. A common structural feature
is the ortho-substituted phenolic group.

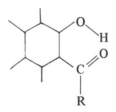

Hydrogen bonding between the OH and the carbonyl oxygen is
believed to play an important role in stabilization. The compound,
t-butyl salol is commonly used in Saran resins where exposure to
sunlight is anticipated.

Many so called "stabilizers" do not actually prevent the degradation of Saran, but merely mask the effects. Dienophiles fall into this
category. Boyer proposed that they function by adding across
conjugated double bonds, thereby destroying the color.[5] The
compounds used for this purpose are normally derivatives of maleic
anhydride.

Antioxidants are not as important in Saran as they are in olefins.
The exact role they play is not known with certainty. They may
serve to prevent oxidation of double bonds formed by dehydrochlorination. The latter process can lead to discoloration. Sterically
hindered phenols are commonly used.

A number of stabilizers function by complexing ferric ion. This includes citrates, phosphates and various amine complexing agents such as ethylene diamine tetraacetic acid.

11.2   MOLDING AND EXTRUSION TECHNOLOGY[4]

The resins used for molding and extrusion are predominantly Saran B containing 10-20% VC. They are formulated with stabilizers, plasticizers and other additives. Formulation is usually required to be able to fabricate the polymers without excessive decomposition. Plasticization is especially important because it lowers the melt temperature and melt viscosity making fabrication in the molten state easier.

Thermal degradation is a serious problem even with formulated resins. The fabrication equipment must be especially designed to minimize polymer hold up. Any polymer that remains for more than a few minutes in a hot zone will decompose. The charred polymer will then flake off and contaminate the fabricated part. Degradation is readily detected by the appearance of black specks in the finished part.

All metal parts that contact the hot molten polymer must be fabricated from corrosion resistant, non-catalytic metals. Nickel alloys such as Duranickel and Hastalloy B are commonly used.

A variety of shapes such as valves and pipe fittings, can be fabricated by injection, compression or transfer molding. The polymer is normally forced into a hot mold to speed up crystallization.

Other shaped articles such as filaments, films, sheet, tubing and pipe are formed by extrusion. Special dies are necessary to handle Saran resins. In extrusion, the hot melt is often quenched to a cold amorphous condition and then allowed to crystallize at a controlled rate in a subsequent heating step.

Orientation can be built into the extrudate by mechanically working the supercooled melt. Monofilaments are uniaxially oriented by stretching. Serdynsky[6] has described the process in detail.

The properties of the filament depend both on the stretch ratio and the temperature of stretching. Some typical properties for a commercial monofilament are listed in Table 11.1.

TABLE 11.1
Typical Properties of 0.010-in. Diameter Round Saran Monofilament

Tensile strength, psi	40,000
Elongation, %	15-25
Water absorption, 27 h immersion	$<0.1\%$
Specific gravity	1.65-1.72
Softening point	115-140 $^\circ$C
Flammability (over 0.050 in.), in./min	Self-extinguishing
Shrinkage, %	Depends upon the formulation and after-treatment (from 2% to 60% is possible)
Heat resistance, intermittent	77 $^\circ$C
Heat resistance, continuous	Shrinks 100 $^\circ$C
Flexural strength, psi	15,000-17,000
Modulus of elasticity, tension, psi $\times 10^2$	0.7-2.0

Some of the uses for VDC copolymer fiber and monofilament include drapery fabric, outdoor furniture and doll hair. These are applications where durability and non-flammability are required.

The most important use for melt fabricated Saran resins is the manufacture of biaxially oriented film. This technology has recently been reviewed by Park[7] and Widiger and Butler.[8] Film is normally made in the bubble process which is illustrated schematically in Figure 11.1.

A molten tube is extruded into a cold bath (2-7 $^\circ$C), then passed into an annealing bath to bring it to the orientation temperature. An air bubble is injected into the tube in the stretching zone. The combination of internal pressure and machine tension causes the tube to blow and stretch biaxially. The blown tube is then collapsed and wound on a roll. The mechanics of biaxial orientation have been analyzed by Alfrey.[9-11] Other aspects of orientation have been considered by Jack.[12] Saran film is used in two ways: as a wrapping material and as shrink film. The wrapping films generally have a lower vinyl chloride content and less plasticizer than shrink film. The film after orientation is slit into the desired widths and

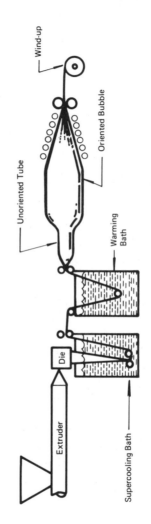

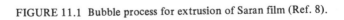

FIGURE 11.1　Bubble process for extrusion of Saran film (Ref. 8).

rewound. The major uses are as household wrap, and as an industrial packaging film particularly for perishable foods.

Biaxially oriented Saran films when heated will tend to shrink before they melt. This property can be optimized to make a heat shrinkable packaging film.[13] The tubular film is simultaneously sealed and cut across the web to make a bag.[8] The bag is loaded, evacuated, tied off at the open end and shrunk in a steam oven. The film draws down tightly over the contents. This method is particularly useful for packaging irregularly shaped objects such as meat. Some typical properties of a shrink film are listed in Table 11.2.

TABLE 11.2
General Physical Properties[a] of "Cryovac S" Film (Ref. 8)

Property	Units	MS 750[b] and MS 755[c] Clear	S–900
Density	g/c^3	1.68	1.61-1.64
Ultimate tensile	psi	10,000-15,000	7,000-14,000
Ultimate elongation	%	30-60	50-100
Modulus of elasticity			
at 73 °F	psi	50,000-65,000	16,000-24,000
at 10 °F	psi	700,000-800,000	
Unrestrained shrink			
at 200 °F	%	25-35	40-50 (205 °F)
at 220 °F	%	34-40	
at 240 °F	%	40-50	
at 260 °F	%	50-57	
Shrink tension			
at 200 °F	psi	190-280	50-200
at 220 °F	psi	180-230	(205 °F)
at 240 °F	psi	140-230	
at 260 °F	psi	90-180	
Shrink temperature, Air	°F		225-260
Shrink temperature, Water	°F		195-205
Heat-seal range (overlap seal)	°F		225-262
Haze	%		1.5-50

cont . . .

Property	Units	MS 750[b] and MS 755[c] Clear	S-900
Water-vapor transmission rate at 100 °F, 100% rh	g/mil-100 in.^{2}-24h	0.1-0.2	0.55
Oxygen permeability	ml/mil-100 in.^{2}-24 h-atm	1.0-1.6	1.6-12.9
Resistance to acids and alkalis		Excellent except to ammonia	
Resistance to grease and oil		Excellent	
Yield	in.^{2}/lb-mil		16,700-17,000
Water absorption			Negligible

[a]   Typical average value.
[b]   MS 750 is designed for hand-wrap applications.
[c]   MS 755 is designed for machine applications.

A relatively new area of application for Saran resins is multi-layer film and sheet. Layers of Saran and other polymers are co-extruded or laminated together to give composites with unique properties. Saran is normally incorporated to provide low permeability. A commercial system is described in Reference 8.

11.3   LACQUER RESINS[14]

The properties of Saran resins make them especially useful as coatings. Coatings are applied to the substrate from lacquer solutions containing 10-20% of a soluble copolymer. Single coats must be thin (~ 0.1 m) in order to remove the solvent efficiently.

The solvents commonly employed are mixtures of toluene with a more polar component. The latter includes THF, MEK, ethyl acetate, cyclohexanone and DMF. The selection of solvent is determined by the composition of the resin.

Lacquer resins usually contain 80 to 90% VDC with acrylonitrile being the other major component.[4] The polymer is further

modified by incorporation of small amounts of the lower alkyl
acrylates and methacrylates. The comonomers are balanced to get a
resin with the desired sealability and solubility. Resins containing
around 80% VDC have relatively low crystallinity and, hence,
are soluble in MEK mixtures. Resins with higher percentages of VDC
dissolve only in the THF mixtures, at ambient temperatures.

Solubility is also strongly influenced by composition distribution. The high VDC fraction in a broad composition distribution
resin tends to precipitate from solution and gel the lacquer. Therefore, lacquer resins are commonly made in a process that yields
narrow distributions.

The major application for VDC copolymer lacquers is the coating
of cellophane.[7] Adhesion of the coating to the substrate was a
major problem in this use. It was overcome by incorporating small
amounts on a vinyl acid such as itaconic acid in the resin.[15] Cellophane coatings are usually formulated with wax and talc to give
slip and antiblock properties to the final product.[16]

Coatings are applied by roller, or dip coating with doctoring by
knife or wire wound rod.[17] The solvent is removed by hot air or
infrared heat at ~ 100 °C. The final heating stage is critical in getting
the coating to crystallize before wind up. The optimum crystallization temperature for most lacquer resins is around 80 °C. The final
product is obtained by rehumidifying the cellophane layer. The
major uses of Saran coated cellophane are in packaging perishable
products.

11.4 LATEXES

VDC copolymer latexes for barrier coatings are a recent innovation
in the packaging fields.[18] The main impetus for development was
the need for a high performance coating system that did not have
the hazards and pollution control problems of lacquer systems. The
latexes now available can deposit a coating similar in performance
to a lacquer coating but from an aqueous medium.

Many VDC copolymers are prepared in emulsion but are not
suitable as coatings latexes because of poor colloidal stability and
lack of film forming ability. The latexes when freshly prepared are
film formers. But, if the VDC level in the polymer is high, they

crystallize in the dispersed state shortly after preparation. The hard crystalline latex particles can no longer coalesce into a continuous film.

Since crystallinity in the final product is very often desirable in barrier coatings a major developmental problem has been to prevent crystallization in the latex during storage but to induce rapid crystallization of the polymer after coating. Gibbs and co-workers[19] found that latexes made with small particle size (1000-1200 Å) could be maintained in an amorphous condition for longer periods. The latexes were prepared by a continuous monomer addition technique to achieve narrower composition distribution. Then, if the VDC content was kept below a critical level (90-92% VDC, depending on comonomer type), the latexes were found to have adequate shelf life together with the ability to crystallize rapidly when dried into a film.

Colloidal stability was achieved initially by using large amounts of a conventional soap. Latexes of this type work well on paper[20] but do not have adequate adhesion to plastic film substrates. These applications require low soap latexes[21,22] or latexes stabilized with comonomeric emulsifiers such as sodium 2-sulfoethyl methacrylate.[23] Latexes of this type are now commercially available.

The major problem in coating polyolefin film is to get adhesion to the substrate. This normally requires corona treatment of the film and proper formulation of the latex to get good wetting.

Latexes are also formulated with antiblock and slip agents. They can be deposited in conventional coating processes such as dip coating followed by a doctor blade. Coating speeds in excess of 1000 ft/min can be attained. Latex coatings are preferably dried in radiant heat ovens. Two coats are normally applied to reduce pinholing.

Latexes intended for barrier coatings usually contain about 90% VDC. Where low heat sealing temperatures are required, the VDC content is usually around 80%. In either case, the comonomers are normally acrylonitrile and the lower alkyl acrylates and methacrylates.

VDC copolymer latexes are also used as binders in paint[24] or to form intumescent coatings.[25] The high chlorine content is an advantage in these applications.

The use of Saran latexes in modifying mortar and concrete is another important application. Relatively low levels of latex significantly upgrade the strength properties of Portland cements in a variety of applications. Some typical test results are shown in Table 11.3.[26]

TABLE 11.3

Effect of Dow Latex on Mortar Properties (Ref. 26)

Test[a]	Value, psi (% Latex based on cement)			
	0	15	20	25
Shear bond strength	50-200	>650	>650	>650
Compressive strength	4480	6180	8430	9670
Tensile strength	380	820	910	1000
Flexural strength	820	1750	1820	1900
Elastic modulus $\times 10^{-6}$	3.40	–	2.52	2.25

[a] All samples aged 28 days at 73 °F, 50% RH before test.

The mechanism by which latex additives reinforce Portland cement is still under investigation, but a hypothesis advanced recently by Isenburg and Vanderhoff[27] seems to account for most observations. They suggest that the lower water/cement ratio achievable in latex modified systems is one important factor. This is augmented by the ability of the polymer to distribute itself throughout the matrix and bridge microcracks formed during the cure.

REFERENCES

1. C. E. Schildknecht, *Vinyl and Related Polymers*, Wiley, N.Y., Chapter VIII (1952).
2. J. F. Gabbett and W. M. Smith, in G. Ham (ed.), *Copolymerization*, Interscience, N.Y., Chapter V (1964).
3. F. Chevassus and R. De Broutelles, *The Stabilization of Polyvinyl Chloride*, (Eng. Trans.), St. Martin's Press, N.Y., Part II (1963).
4. R. A. Wessling and F. G. Edwards, *Encyc. Polym. Sci. and Tech.*, **14**, 540 (1971).

5. R. F. Boyer, *J. Phys. Coll. Chem.*, **51**, 80 (1947).
6. E. D. Serdynsky, in H. F. Mark, S. M. Atlas and E. Cernia (eds`, *Man-Made Fibers*, Vol. III, Interscience, N.Y., p. 303 (1968).
7. W. R. R. Park (ed.), *Plastics Film Technology*, Van Nostrand Reinhold Co., N.Y. (1969).
8. A. T. Widiger and R. L. Butler, in O. J. Sweeting (ed.), *The Science and Technology of Polymer Films*, Vol. II, Interscience, N.Y., Chapter 6 (1971).
9. T. Alfrey, Jr., *SPE Trans.*, **5**, 68 (1965).
10. R. A. Wessling and T. Alfrey, Jr., *Trans. Soc. Rheol.*, **8**, 85 (1964).
11. T. Alfrey, Jr., *Appl. Polym. Symp.*, **17**, 17 (1971).
12. J. Jack, *Brit. Plastics*, **34**, 312 (1961); *ibid.*, **34**, 391 (1961).
13. G. V. Moore and G. R. Irons, U.S. 2,538,025, to The Dow Chemical Company (1951).
14. *Saran F Resins*, Technical Bulletin, The Dow Chemical Company Europe (1969).
15. G. Pitzel, U.S. 2,570,478, to du Pont.
16. D. K. Owens, in O. J. Sweeting (ed.), *The Science and Technology of Plastic Films*, Vol. 1, Interscience, N.Y., Chapter 9 (1968).
17. H. H. Sineath and W. A. Pavelchek, *ibid.*, Vol. II, Chapter 7.
18. R. F. Avery and R. K. von Leer, *Mod. Packaging*, **39**, 36 (1965).
19. D. S. Gibbs, The Dow Chemical Company, unpublished results.
20. R. F. Avery, *Tappi*, **45**, 356 (1962).
21. D. G. Grenley, U.S. 3,353,992 (1967), to The Dow Chemical Company.
22. D. K. Owens, *J. Appl. Polym. Sci.*, **14**, 1725 (1970).
23. D. S. Gibbs and R. A. Wessling, U.S. 3,617,368, to The Dow Chemical Company (1971).
24. J. C. Bax, *J. Oil Col. Chem. Assoc.*, **53**, 592 (1970).
25. A. R. French, *Paint Manu.*, **40**, 57 (1970).
26. The Dow Chemical Company, Product Bulletin No. 191-18-70 (1968).
27. J. E. Isenburg and J. W. Vanderhoff, *Polymer Preprints*, **14**, 1197 (1973).

Subject Index

185

Author Index